# POLLUTION

# POLLUTION

P. R. Trivedi

**APH PUBLISHING CORPORATION**
4435-36/7, Ansari Road, Darya Ganj
New Delhi-110 002

*Published by*

S.B. Nangia
**APH Publishing Corporation**
4435-36/7, Ansari Road, Darya Ganj
New Delhi-110 002
PH. : 23274050
Email : aphbooks@gmail.com

2026



*Typesetting at*

**Rawat Computers**
Gandhi Nagar, Delhi

*Printed at*

**Balaji Offset**
Navin Shahdara, Delhi-32

# Preface

Man's capability to transform his environment can bring the benefits of economic development and an opportunity to enhance the quality of life. But this same power, incorrectly applied, can also cause incalculable harm to the natural environment and consequently to human life. There should be no delay in tackling the task of solving these environmental problems, as these problems have a cumulative impact. Delayed remedial action will cost considerably more and the damage will become irreversible. There is only one world to pollute; if this is ruined, there is no other. Man will survive only as long as the Earth survives.

The perception that man lives in close relationship with all living things, plants and animals, is quite old. Ancient and medieval philosophers have often emphasized the inter-relatedness of the entire living world. Recent developments in science have not only confirmed this view but have gone beyond to assert the interdependence of all life, microbial, plant and animal, between themselves and with the physical environment of land, air, water and solar energy, this interdependence is embedded in the evolution of life during the past three billion years or more with the origin of various forms of life, the extinction of others, and the formation of our atmosphere, water bodies and land masses, these ideas have been sharpened in the last nearly fifty years by the concept of ecosystem which defines an area in which all living creatures and the soil-water-air-environment of this physical region of the earth have developed various inter-relations to produce a quasi-stable or almost stable system of mutual independence.

The primary goal of preparing these volumes is to present a balanced view of the diversity of issues that relate to the environment and to provide the fundamental information that will allow readers to understand the complexities of those issues. The other goal for preparing these volumes is to focus on underlying principles and the exciting advances in understanding that have characterised the last decade of environmental science. This approach will help each reader establish a base of general knowledge about environmental issues from which rational, educated political choices and opinions can be fashioned independently.

*Editor*

# Contents

# 1

# Air Pollution

The pollution of atmosphere is not a new phenomenon. Air pollution existed even before man became aware of nature, though man made pollution can be traced historically, to the discovery of the fire by the caveman. It has however, assumed menacing proportions recently due to rapid increase in population growth and Industrialization

In recent years, great concern has been universally voiced regarding environmental pollution arising as a side effect of industrial activities. In our national situation with the growth of industries, more and more toxic substances are either used as raw materials or given off during manufacturing processes in the form of dust, fumes, vapours and gases. These pollutants ultimately dissipate in the working environment and pose occupational health hazards. According to a U.S. Estimate, there are some half a million man made substances already present as pollutants in our total environment. These pollutants, when get into the air and water, disturb the ecological balance of nature. Control of air contaminants is therefore essential for we are just the custodians of environment for our future generation,. This can be achieved by properly recognising, evaluating and controlling the industrial environment.

Considering the above, in India, may well be

satisfied with consciousness in this matter and it may be mentioned that several Air and Environmental Authorities have been in existence from the 1960-65 period onwards. An Indian 'Water Pollution Bill' came into force from 1969 and a National Committee on Environmental behaviour started functioning from the early seventies. As a result of increasing awareness, four more acts came into existence which are the water (prevention and control of pollution act, 1974, the air (prevention and control of pollution) cess act 1977, the air (prevention and control of pollution act, 1981) and the environment protection act, 1986.

The dry air contains 78.09% nitrogen by volume and 20.94% oxygen the remaining 0.97% is composed of gaseous mixture of carbon dioxide, helium, argon, krypton, nitrous oxide and xenon, as well as very small amounts of some other organic and inorganic gases whose amount in the atmosphere vary with time and place. Various amounts of contaminants continuously enter the atmosphere through both natural and man-made processes that exist upon the earth. That portion of these substances which interacts with the environment to cause toxicity, disease, aesthetic distress, physiological effects or environmental decay, has been labelled by man as a 'pollutant'. In general, the actions of people are the primary cause of pollution and as the population increases, the attendant pollution problems also increase proportionately. The first significant Change in man's effect on nature came with his discovery of fire. Prehistoric man built a fire in his cave for cooking, heating and to provide light. The problem of air pollution came into existence at this time.

The British Parliament passed an Act in 1273, forbidding the burning of coal in London because it was beginning to choke the atmosphere. In 1300, King

Edward I issued a royal proclamation, "Whoseever shall be found guilty of burning coal all suffer the loss of his head." In 1306, a man was executed for violating this regulation. Later, the law fell into disuse as the industrial revolution took place in England.

Air pollution means different things to different people. To the householder it may be an eye irritation and soiled clothing, to the farmer damaged vegetation, to the pilot dangerously reduced visibility and to industries problems of process control and public relations.

Further, the problem of air pollution varies from place to place. For example, air pollution in Tokyo is not the same as that in Bombay.

Today, the natural terrain that surrounds large cities is recognized as having a significant bearing on the problem of air pollution. However, this is not an altogether new concept either. Historians tell us that Los Angeles which in recent years has become a national symbol of comparison for excessive smog levels, was known as the 'valley of smokes' when the Saniards first arrived there.

It has been found that a significantly increasing volume of particulate matter entering the atmosphere, scatters the incoming sunlight. This reduces the amount of heat that reaches the earth and tends to reduce its temperature. This decreasing mean global temperature of recent years have been attributed to the rising concentration of air-borne particles in the atmosphere. A counter acting phenomenon commonly referred to as the "green-house effect" is caused by the increasing amount of carbon dioxide found in the atmosphere. It has been estimated that if carbon dioxide content in the atmosphere generated in combustion processes continues to increase at the present rate, the mean global temperature could rise by 4°C in the next five decades. There has been conjecture that this might become a matter of great importance because small temperature increase could cause a partial melting of the ice caps of the earth causing continental flooding and devastating effects on man.

Air pollution can cause death, impair health, reduce visibility, bring about vast economic losses and contribute to the general deterioration of both our cities and country-side. It can also cause intangible losses to historical monuments such as the Taj Mahal which is believed to be badly affected by air pollution.

On account of large scale industrial activities in advanced countries notably the USA, UK, and other European countries, fall of acid rain has been reported in Scandinavia. This has reduced forest growth in Scandinavia. In Canada, thousands of lakes have been destroyed due to acid rain. Apart from the international issues involved, the basic ecology is affected. Large scale deforestation apart from creating an imbalance in the oxygen proportion of the atmosphere, affects weather and rain patterns as well. Industrial activity, particularly in thermal power stations, cement plants, oil refineries, chemical complexes, metallurgical industries, steel plants and fertilizer complexes, causes major, problems of air pollution. The effects of air pollution are felt more by the elderly and infants as chest and respiratory complaints are very common among them. Tragic instances of death have also been reported in air pollution episodes such as the London smog of 1952, and the Bhopal gas tragedy of 1984.

It is, therefore, a matter of great importance that engineers of all disciplines consciously incorporate in their designs sufficient constraints and safeguards to ensure that they do not contribute to atmospheric pollution, in addition, they must apply their ingenuity and problem-solving abilities to eliminate air pollution where it exists and restoring the natural environment.

There are three methods of identifying air pollution:

1. Sensory recognition
2. Physical measurements of pollution.
3. Effects on plants, animals, and buildings.

These methods are widely used and they have their merits and demerits.

1. Sensory recognition—Usually the first awareness of an air pollution problems is through some effects on the individual. These are:

(a) strong or unusual odours

(b) Reduction in visibility

(c) Eye irritation

(d) Acid taste in the mouth

(e) Feel of grit under foot

There are highly subjective phenomena and vary from individual.

(2) Physical measurement—While sensory perception may provide the first indication of the presence of most of the contaminants in the air, it is often not possible to detect trace quantities of many air-borne toxic substances or the presence of radioactive matter through the senses. Their identification required physical measurement by standard methods of sampling and analysis.

(3) Effects on plants, animals and buildings—Effects of air pollution can be observed on the growth of plants and health of animals. Similarly, its deleterious effect on building can also be observed. Thus plants, animals, and buildings act to some extent as indicators of certain atmospheric impurities.

**Classification of pollutants**

Air pollutants can be classified as follows:

1. Natural contaminants
   e.g., natural fog, pollen grains, bacteria, and products of volcanic eruption
2. Aerosols (particulates)
   e.g., dust, smoke, mists, fog and fumes.
3. Gases and vapours

The various gases and vapors, which are important air contaminants are given in Table 1.

**Table 1:** *Air Contaminants*

| *No.* | *Group* | *Examples* |
|---|---|---|
| 1. | Sulphur compounds | $SO_2$, $SO_2$, $H_2S$, mercaptans |
| 2. | Nitrogen compounds | NO, $NO_2$, $NH_3$ |
| 3. | Oxygen compounds | $O_3$, CO, $CO_2$ |
| 4. | Halogen compounds | HF, HCl |
| 5. | Organic compounds | Aldehydes, hydrocarbons |
| 6. | Radioactive compounds | Radioactive gases |

In addition to the above, studies by various investigators have indicated, that some of these contaminants undergo chemical reactions when they enter the atmosphere. As a result the end products formed are more harmful than the original contaminants. For example, it is reported that unsaturated hydrocarbons react with nitrogen dioxide in sunlight to form smog.

**Natural contaminants**

Among natural contaminants pollen is important because of its peculiar properties irritating to some individuals. Pollen grains are the male gametophytes of gymnosperms and angiosperms and they are discharged into the atmosphere from weeds, grasses and trees. Because of wind pollination, thousands of pollen grains are liberated. While air transported pollen grains range chiefly between 10 and 50μ (micron) in size, some have been found to be as small as 5μ and as large as 100μ in diameter.

From the point of view of pollution, air pollutants are significant because of the allergic responses produced in sensitive individuals. Many people suffer from asthma or hay fever. While most victims have an uncomplicated type of hay fever in which the symptoms disappear at the end of the pollen season, some develop bronchitis, bronchial asthma, and dermatitis.

**Aerosols**

Aerosols refer to the dispersion of solid or liquid particles of microscopic size in gaseous media, such as dust, smoke, or mist. An aerosol can also be defined as a colloidal system in which the dispersion medium is a gas and the disappeared phase is solid or liquid. The term 'aerosol' is used during the time it is suspended in the air. After it has settled either by virtue of its weight, by agglomeration, or by impact on a solid or liquid surface, the term no longer applies. Thus, particulate matter is an air pollutant only when it is an aerosol. However, it is a nuisance both as an aerosol (visibility reduction) and as settled or deposited matter (soiling of surfaces, corrosions). Aerosols differ widely in terms of particle size, particle density and their importance as pollutants. Their diameters generally range from 0.01µ or less, up to about 100µ. The following are the various aerosols.

**Dust**

Dust is made up of solid particles predominantly larger than those found in colloids and capable of temporary suspension in air or other gases. They do not tend to flocculate except under electrostatic forces; they also do not diffuse but settle under the influence of gravity. Dust is produced by the crushing, grinding etc., of organic and inorganic materials. Generally they are over 20µ in diameter, although some are smaller. Fly ash from chimneys varies from 80-3µ; cement, from 150-10µ; foundry dust, from 200-1µ. Most of the dust particles settle to the ground as dust fall, but particles 5µ or smaller tend to form stable suspensions.

**Smoke**

Smoke consists of finely divided particles produced by incomplete combustion. It consists predominantly of carbon particles and other combustible materials. Generally the size of the particles is less than 1µ. The size of coal smoke particles range from 0.2-0.01µ and oil smoke particles from 1.0-0.03µ.

**Mists**

This term refers to a low concentration dispersion of liquid particles of large size. In meteorology, it means a light dispersion of minute water droplets suspended in the atmosphere. Natural mist particles formed from water vapour in the atmosphere are rather large, ranging from 500-40µ in size. The particles may coalesce.

**Fog**

Fog refers to visible aerosols in which the dispersed phase is liquid. Formation by condensation is usually implied. In meteorology, it refers to dispersion of water or ice in the atmosphere near the earth's surface reducing visibility to less than ½ km. In natural fog the size of the particles ranges from 40-1.0μ.

**Fumes**

These are solid particles generated by condensation from the gaseous state generally after volatilization from melted substances, and often accompanied by a chemical reaction such as oxidation. Fumes flocculate and sometimes coalesce.

Table 1 indicates the major sources of atmospheric dust.

**Table 1:** *Sources of Atmospheric Dust*

| | *Sources* | *Examples* |
|---|---|---|
| 1. | Combustion | Fuel burning (coal, wood, fuel oil) incineration (house and municipal garbage) Others (open fires, forest fires, tobacco smoking) |
| 2. | Materials handling and processing | Loading and unloading (sand, gravel, coal, ores, lime, cement) Crushing and grinding (ores, stone, cement, rocks, chemicals) Mixing and packaging (chemicals, fertilisers) Food processing (flour, cornstarch, grains) Cutting and forming (saw mills, wall board, plastics) Metallurgical (foundries, smelters) Industrial (paper, textiles manufacture) |
| 3. | Earth-moving operations | Construction (road, |

| | | |
|---|---|---|
| | | buildings, dams, site clearance) Mining (blasting) Agriculture (soil filling, land preparation) Winds |
| 4. | Miscellaneous | House cleaning, Mud road cleaning, Crop spraying, Poultry feeding Engine, exhaust |

**Gases**

Following are the gases in air pollutants:

**Sulphur Dioxide**

This is one of the principal constituents of air pollutants. The main source of sulphur dioxide is the combustion of fuels, especially coal. Therefore, its concentration in the atmosphere depends upon the sulphur content of the fuel used for heating and power generation. The sulphur content of fuels varies from less than 1% for good quality anthracite to over 4% for bituminous coal. In the USA in recent years, there has been a progressive decrease in the average atmospheric concentrations of sulphur dioxide, of several cities, owing to the increased use of coal with a low sulphur content. Air pollution control measures in St Louis prohibit the use of high-sulphur coal.

Most crude petroleum products contain less than 1% sulphur: a few contain upto 5%. Refining processes tend to concentrate sulphur compounds in the heavier fractions. Fuel gases also contain sulphur, but, in small quantities. About 80% of the sulphur in coal and nearly all that in liquid and gaseous fuels is found in flue gases in the form of sulphur dioxide. The remaining sulphur in coal is present as inorganic sulphur and thus remains in the ash. Generally, the concentration of sulphur dioxide in flue gases ranges from 0.005-0.25% and occasionally is as high as 0.4%.

Another common source of sulphur dioxide in the atmosphere, is metallurgical operations. Many ores, like zinc, copper and lead, are primarily sulphides. During the smelting of these ores, sulphur dioxide is evolved in stack concentrations of 5-10% ($SO_2$) but this, can be recovered in the form of sulphuric acid. Among the miscellaneous operations releasing sulphur dioxide into the atmosphere are sulphuric acid plants and paper manufacturing plants. The quantities are usually low and therefore easily amenable to control measures.

The open burning of refuse and municipal incinerators also contribute some amount of sulphur dioxide to the atmosphere.

**Hydrogen Sulphide**

Hydrogen sulphide is a foul smelling gas. The sources of its natural emission include anaerobic decay processes on land, in marshes and in the oceans. Volcanoes and natural water springs emit hydrogen sulphide to some extent.

One of the major sources of Hydrogen sulphide is the Kraft pulp industry, which uses a sulphide process for manufacturing paper. The other industrial sources of hydrogen sulphide are petroleum refineries, coke-oven plants, viscose rayon plants, and some chemical operations.

Other sulphur compounds that are of interest in air pollution, principally because of their strong odour, are methyl mercaptan ($CH_3$ SH), dimethyl sulphide ($CH_3$ $SCH_3$) dimethyl disulphide ($CH_3$ SS $CH_3$) and their higher molecular homologs. The mercaptans are emitted in mixtures of pollutants from some pulp mills, petroleum refineries, and chemical manufacturing plants.

### Hydrogen Fluoride

The major sources of fluorides are the manufacturers of phosphate fertilizers, the aluminium industry, brick plants, pottery, and ferro-enamel works. Small amounts are also emitted from other metallurgical operations, such as zinc foundries and open-hearth steel furnaces. Small amounts are also liberated in the burning of coal, which normally contains about 0.01% fluorine. Hydrogen Fluoride is an important air contaminant even in extremely low concentrations of 0.001-0.10 ppm by volume. In these concentration levels, hydrogen fluoride is more important in terms of injury to vegetation and animals than in terms of injury to humans. The high degree of toxicity of fluorine compounds renders the control of such emissions imperative for industries manufacturing aluminium and phosphate fertilizers.

### Chlorine and Hydrogen Chloride

Chlorine is found in polluted atmospheres as the element itself (chlorine) as hydrogen chloride, as chlorine containing organic compounds such as perchloroethylene and as inorganic chlorides. The last mentioned compounds are solids and hence found in particulated form; the other materials mentioned are present as gases.

The most common sources of chlorine in the atmosphere are from operations in which it is manufactured or used to produce other chemicals. Also as chlorine is used in water purification plants, in sewage plants and in swimming pools, equipment failure sometimes leads to leakage of chlorine into the atmosphere.

Hydrogen chloride is evolved in numerous industrial chemical processes, but it is so easy to recover, that little reaches the atmosphere.

The main effects of chlorine and its compounds are respiratory irritation from chlorine, corrosion by hydrogen chloride and damage to vegetation from chlorine and hydrogen chloride.

**Oxides of Nitrogen**

It is probable that oxides of nitrogen are the second most abundant atmospheric contaminants in many cities, ranking next to sulphur dioxide. Generally the highest concentration of nitrogen oxides in gaseous emissions occurs in effluents from industries where nitric acid is produced or used in chemical reactions. The next highest concentration is in automobile exhausts. The come effluents major from large power plants, and then to a small extent those from low heat burners and furnaces.

Out of seven oxides of nitrogen ($N_2O$, NO, $NO_2$, $NO_3$, $N_2O_3$, $N_2O_4$, $N_2O_5$), only nitric oxide and nitrogen dioxide arise from many human activities and are classified as pollutants. In atmospheric analyses they are usually reported as 'total oxides of nitrogen' or "$NO_x$". It is a standard practice in the chemical industry to absorb and recover significant quantities of oxides of nitrogen.

Ammonia and ammonium salts are not important air contaminants. Ammonia is an important raw material in fertilizer and organic chemical industries and in the manufacture of nitric acid by the oxidation process; its recovery is a matter of fundamental important in the economical operations of such processes.

**Carbon Monoxide**

Carbon Monoxide, an odourless and colourless gas has its major origin in the incomplete combustion of carbonaceous materials. It is a highly poisonous gas and is generally classified as an asphyxiant. The chief source of carbon monoxide in the atmosphere is combustion, especially due to automobile exhausts. However, except for motor vehicles and other internal combustion engines, very little carbon monoxide is found in the gaseous emissions from properly adjusted, properly operated installations. Although certain industrial operations, such as electric and blast furnaces, some petroleum refining operations, gas manufacturing plants, and coal mines, are potential contributors of carbon monoxide to the atmosphere, automobile exhausts are by far the most important source.

**Ozone**

The origin of ozone that is found in the air has not been clarified, but it is likely that combustion and sunlight are involved in its production. Ozone is poisonous and smelly. It exists in great abundance under natural conditions in the upper atmosphere.

**Aldehydes**

These are produced by the combustion of gasoline, diesel oil, fuel oil, and natural gas. Incomplete oxidation of motor fuel and lubricating oils leads to the formation of aldehydes. Lower aldehydes may be present in the atmosphere in concentrations about as high as sulphur dioxide. They may also be formed in the air because of photochemical reactions. Formaldehyde is irritating to the eyes.

**Organic Vapours**

These contaminants include a large number of chemical compounds, including paraffins, olefins, acetyleness, aromatic hydrocarbons, chlorinated hydrocarbons, etc. They are produced by combustion processes (particularly in automobiles), household incinerators, and petroleum processes. It is also probable that they undergo changes in the atmosphere which contribute to the formation of smog.

**Radioactive Gases**

A major source of radioactive gases and particulate is the nuclear power reactor and related fuel handling facilities. The other sources are experimental accelerators, testing of nuclear bombs in the atmosphere, agricultural, industrial and medical use of radioactive isotopes. Another source of radioactive particulates and gases that is increasing in importance, is the nuclear fuel reprocessing plant.

**Primary and Secondary Air Pollutants**

Air pollutants can also be broadly classified into two general groups—primary air pollutants and secondary air pollutants.

Primary air pollutants are those emitted directly from identifiable sources.

Examples of primary air pollutants:

1. Finer particles (less than 100m in diameter)
2. Coarse particles (greater than 100 in diamcter)
3. Sulphur compounds
4. Oxides of nitrogen
5. Carbon monoxide
6. Halogen compounds
7. Organic compounds
8. Radioactive compounds

Finer aerosols include particles of metal, carbon, tar, resin, pollen, bacteria, etc.

Secondary air pollutants are those which are produced in the air by the interaction among two or more primary pollutants, or by reaction with normal atmospheric constituents, with or without photo-activation. Examples of secondary air pollutants:

1 Ozone

2. Formaldehyde

3. PAN (proxy acetyl nitrate)

4. Photochemical smog

5. Formation of acid mists($H_2SO_4$) due to reaction of sulphur dioxide and dissolved oxygen, when water droplets are present in the atmosphere.

**Smog**

Smog is a synchronym of two words -- smoke and fog. It can be of two types -- photochemical or coal induced.

Photochemical smog is restricted to highly motorised areas in metropolitan cities. e.g., Los Angeles. It occurs under adverse meteorological conditions when the air movement is restricted. Smog is caused by the interaction of some hydrocarbons and oxidants under the influence sunlight giving rise to dangerous peroxy acetyl nitrate (PAN). Its main constituents are nitrogen oxides, PAN, hydrocarbons, carbon monoxide and ozone. It reduces visibility, causes eye irritation, damage to vegetation and cracking of rubber.

The fog from burning coal covers urban areas at night or on cold days when the temperature is below $10^0$C and when calm meteorological conditions prevail, e.g., London (December 1952). This fog consists of smoke, sulphur compounds and fly ash. Prolonged exposure for smog may result in a high mortality rate especially among the elderly and those who have histories of Chronic bronchitis, asthma, bronchopneumonia and other lung or heart diseases.

**Stationary and Mobile Sources**

Emissions (air pollutants) may be classified by source, as stationary or mobile. Another method of classifying emission sources is by:

1. Point sources (large stationary sources)
2. Area sources (small stationary sources and mobile sources with indefinite routes)
3. Line sources (mobile sources with definite routes)

The following chart gives a picture of emission inventory sources classification.

**Air pollution Due to Automobiles**

This section discusses the contribution of automotive emissions in aggravating the air pollution menace. Various causes of the genesis and exodus of these pollutants have been identified and methods to control them have been outlined. Certain modifications in the engine design and operating variables are suggested.

The three main types of automotive vehicles being used in our country are(a) passenger cars powered by four stroke gasoline engines, (b) motor cycles, scooters and autorickshaws powered mostly by small two stroke gasoline engines, and (c) large buses and trucks powered mostly by four stroke diesel engines

Emissions from gasoline powered vehicles are generally classified as :(a) Exhaust emissions, (b) Crank-case emissions and, (c) Evaporative emissions. The amount of pollutants, that an automobile emits depends on a number of factors, including the design and operation. Of the hydrocarbons emitted by a car with no controls, the exhaust gases account for roughly 65%, evaporation from the fuel tank and carburettor for roughly 15% and blowby or crankcase emission (gases that escape around the piston rings) for about 20%. Carbon monoxide, nitrogen oxides and lead compounds are emitted almost exclusively in the exhausts gases. Table 4 gives the emissions from a typical Indian car at different conditions.

Diesel powered vehicles create relatively minor pollution problems compared to gasoline powered ones. The diesel engines exhausts only about a tenth of the amount of carbon monoxide exhausted by a gasoline engine, although its hydrocarbon emissions may be more than those of the gasoline engine. Blowby is negligible in the diesel, since the cylinders contain only air on the compression stroke. Evaporative emissions are also low because the diesel engine uses a closed injection fuel system and because the fuel is less volatile than gasoline.

**Air pollution from major Industrial Operations**

Even though there are many different sources which contribute to air pollution, industries contribute a major share. There are a number of industries which are responsible for air pollution, but it is not possible to cover all of them. Table 5 shows the specific pollutants found in the emissions of some typical industrial and other sources.

**Effects of air pollution on human Health**

Air pollution is one of the greatest environmental evils. The air we breathe has not only life-supporting properties but also life-damaging properties. Under ideal conditions the air we inhale has a qualitative balance that maintains the well-being of man. But when the balance among the air components is disturbed, or in other words, if it is polluted, it may affect human health. An average man breathes 22,000 times a day and takes in 16 kg of air each day. It far exceeds the consumption of food and water. It has been estimated, that a man can live for five weeks without food and five days without water, but only for five minutes without air.

All the impurities in the inhaled air do not necessarily cause harm. Depending upon the chemical nature of the pollutants, some may be harmful when present in the air in small concentrations and others only if they are present in high concentrations. The duration of exposure of the body to polluted air is also an important factor. Therefore, the prime factors affecting human health are:

1. Nature of the pollutants
2. Concentration of the pollutants
3. Duration of exposure
4. State of health of the receptor
5. Age group of the receptor

Generally speaking susceptibility to the effects of air pollution is great among infants, the elderly, and the infirm. Those with chronic diseases of the lungs or heart are thought to be at great risk. Pre-school and school children appear to be both sensitive and specifically reactive to air pollution health effects. Another point to be noted is that the effect of air pollution on human health is worst during the winter season, when pollution levels reach a climax.

An objectionable odour, visibility reduction, eye irritation or vegetation damage are useful guides to the likelihood or severity of health effects. A grey ball over a city or an industrial area can have a depressing effect and impair the enjoyment of life. There is no doubt, however, that the urgency with which steps are taken to improve air quality will depend very much on how serious the risk of ill health from air pollution is thought to be.

**Table 2.** *Air Pollution Episodes*

| S.No. | Month and year | Place | Mortality |
|---|---|---|---|
| 1. | December, 1930 | Meuse Valley (Belgium) | 63 |
| 2. | October, 1948 | Donora (Pennsylvania) | 20 |
| 3. | November, 1950 | Poza Rice (Mexico) | 22 |
| 4. | December, 1952 | London | 4000 |
| 5. | November, 1953 | New York | 220 |
| 6. | January, 1956 | London | 1000 |
| 7. | December, 1957 | London | 750 |
| 8. | December, 1962 | London | 700 |
| 9. | January, 1963 | New York | 300 |
| 10. | November, 1966 | New York | 168 |
| 11. | December, 1984 | Bhopal (India) | 2000 |

**Health Effects**

1. Eye irritation
2. Nose and throat irritation.
3. Irritation of the respiratory tract
4. Gases like hydrogen sulphide, ammonia and mercaptans cause odour nuisance even at low concentrations.
5. Increase in mortality rate and morbidity rate.
6. A variety of particulates particularly pollens, initiate asthmatic attacks.
7. Chronic pulmonary diseases like bronchitis and asthma, are aggravated by a high concentration of

$SO_2$, $NO_2$, particulate matter and photochemical smog.

8. Carbon monoxide combines with the haemoglobin in the blood and consequently increases stress on those suffering from cardiovascular and pulmonary diseases.
9. Hydrogen fluoride causes diseases of the bone (fluorosis), and mottling of teeth.
10. Carcinogenic agents cause cancer.
11. Dust particles cause respiratory diseases. Diseases like silicosis, asbestosis, etc., result from specific dusts.
12. Certain heavy metals like lead may enter the body through the lungs and cause poisoning.

## HEALTH EFFECTS OF SPECIFIC POLLUTANTS

### Sulphur Dioxide

Sulphur dioxide is an irritant gas which affects the mucous membranes when inhaled. Under certain conditions, some of the air-borne sulphur dioxide gas is oxidized to sulphur trioxide. Each of these two gases in the presence of water vapour or water, forms sulphurous and sulphuric acid respectively. Sulphur trioxide is a very strong irritant, much stronger than sulphur dioxide, causing severe bronchospasms at relatively low levels of concentration.

### Carbon Monoxide

Carbon monoxide has a strong affinity for combining with the hemoglobin of the blood to form carboxyhaemoglobin, COHb. This reduces the ability of the haemoglobin to carry oxygen to the body tissues. CO has about two hundred times the affinity of oxygen for attaching itself to the haemoglobin, so that low levels of CO can still result in high levels of COHb.

Carbonmonoxide also effects the central nervous system. It also responsible for heart attacks and a high mortality rate.

**Oxides of Nitrogen**

Of the seven oxides of nitrogen known to exist in the ambient air, only two are thought to affect human health. These are nitric oxide (NO) and nitrogen dioxide ($NO_2$). While some questions remain about haemoglobin reactions with oxides of nitrogen, there is no positive evidence that nitric oxide exposure is a health hazard associated with community air pollution.

Nitrogen dioxide is known to cause occupational disease. Among occupations with $NO_2$ hazards are the manufacture of nitric acid, exposures of farmers to silage that has high nitrate fertilisation, electric welding, and mining utilizing nitrogen compounds as explosives. It is estimated that eye and nasal irritation will be observed after exposure to about 15ppm of nitrogen dioxide and pulmonary discomfort after brief exposures to 25 ppm of nitrogen dioxide.

**Hydrogen Sulphide and Mercaptans**

Hydrogen sulphide is a. foul smelling gas. It is well known for its rotten egg like odour. Exposures to hydrogen sulphide for short periods can result in fatigue of the sense of smell.

Other sulphur compounds that are of interest in air pollution mainly because of their strong odours are methyl mercaptan ($CH_3SH$) and ethyl mercaptan ($C_2H_5SH$). But it has been reported that at the concentrations at which they are odour nuisances, they have no other effect on human health. In fact, mercaptans are often added to natural or manufactured gas supplies so that leakage of gas will be noticed.

**Ozone**

Ozone is a gas that has an irritant action in the respiratory tract, reaching much deeper into the lungs than the oxides of sulphur.

**Fluorides**

Fluorides present in air, range from those which are extremely irritant and corrosive like hydrogen fluoride to relatively non-reactive compounds. But, fluorine is a cumulative poison even under condition of prolonged exposure and in sub-acute concentrations.

**Lead**

The main source of lead in urban atmosphere is the automobile. It creates urban concentrations of inorganic lead of about 1-3mg/m$^3$, with high values in area of heavy traffic. Inorganic lead acts as an agent which causes a variety of human health disorders. The effects include gastro-intestinal damage, liver and kidney damage, abnormalities in fertility and pregnancy, and mental development of children gets affected.

**Hydrocarbon Vapours**

Some of the hydrocarbon vapours in the atmosphere have health implications. The effect of formaldehyde is primarily irritating. It is a major contributor to eye and respiratory irritation caused by Photochemical smog.

**Carcinogenic Agents**

Carcinogenic agents are responsible for cancer. For example, the poly-cyclic organic compound, 3, 4-benzpyrene. The origin of these compounds is in incomplete combustion of hydrocarbons and other carbonaceous materials. They are also reported to be present in exhaust discharges from I.C. engines. In addition to poly-cyclin organic compounds, it has been found that some aliphatic hydrocarbons are also carcinogenic.

**Insecticides**

Insecticides are not only harmful for insects but also poisonous for man, e.g., DDT (Dichloro diphenyl trichloroethane). They can affect the central nervous system and may attack other vital organs. In fact, DDT has been found in mother's milk in western countries and even in our own country. Hence, the use of DDT has been banned in the USA. The United Nations Environment Programme (UNEP) has warned against the undesirable effects of indoor spraying of DDT. It has also been reported that indoor spraying affects domestic livestock.

According to a study conducted at the Industrial Toxicology Research Centre, Lucknow (India), the accumulation of pesticides in the environment due to their growing use for agricultural purposes can also cause premature labour and abortion, due to high concentration of pesticides in the body of expectant mothers.

**Radioactive Isotopes**

The important radioactive isotopes that may reach ambient air are lodine 131, phospourous 32, Cobalt 60, Strontium 90, radium 226, Carbon 14, Sulphur 35, Calcium 45 and Uranium. The major sources of radioactive air pollutants are:

(a) Nuclear reactors

(b) Experimental accelerators

(c) Scientific and medical use of radioactive isotopes

(d) Agricultural and industrial use of radioactive isotopes as tracers, and

(e) Testing of nuclear bombs in the atmosphere.

The serious health effects are anaemia, Leukaemia and cancer. Radioactive isotopes also cause genetic defects and seterility, as well as embryo defects and congenital malformations. It also shortens the life span of an individual.

**Effects of Air Pollution on Animals**

Interest in the effects of air pollution on animals has generally developed as a corollary to the concern about its influence on human health. Most of the information concerning the natural exposure of animals to air pollution is contained in the reports of some major air pollution disasters e.g., Donora, London and Poza Rica. Recently, considerable information has been reported from medical research laboratories which describes the results of experimental exposure of small animals to varoius air pollutants Animals used for laboratory work were mice, rabbits, rats, guinea pigs and monkeys.

**Air pollution Effects on Farm Animals**

The process by which farm animals get poisoned is entirely different from that by which human beings exposed to polluted atmospheres are poisoned. In case of farm animals it is a two-step process.

1 Accumulation of the air - borne contaminant in the vegetation and forage

2. Subsequent poisoning of the animals when they eat the contaminated vegetation.

In case of human beings working in polluted atmospheres in factories, the concern is for the harmful substances that are directly inhaled, whereas in the case of the farm animals, the danger obviously is not in inhaling the polluted air, but rather than ingestion of forage which has been contaminated with pollutants like fluorine from the air. The three pollutants responsible for most livestock damage are fluorine, arsenic and lead. These pollutants originate from industrial sources or from dusting and spraying.

**Fluorine**

Of all farm animals, cattle and sheep are the most susceptible to fluorine toxicosis. Horses appear to be quite resistant to fluorine poisoning. Poultry are probably the most resistant to fluorine, of all farm animals.

**Symptoms of Acute Fluorine Poisoning**

Lack of appetite, rapid loss in weight, decline in health and vigour, lameness, periodic diarrhoea, muscular weakness and death, characterize the acute form of fluorine poisoning. It may also result in considerable increase of bone fluorine. But acute poisoning due to fluorides is unlikely in a majority of the cases.

**Symptoms of Chronic Fluorine Poisoning**

Fluorine is a cumulative poison under conditions of continuous exposure to sub-acute doses. Fluorine is also a protoplasmic poison. It has a marked affinity for calcium and interferes with normal calcification. Animals have been reported to be more resistant than humans to dental mottling. Cattle and sheep are the most frequently affected animals. Teeth in the process of formation are easily affected. Hence tooth symptoms are a sensitive and unique criterion of chronic fluorosis. Excessive wearing of incisor and molar teeth may occur at high levels of fluorine intake.

Bone lesions may develop at any age. A bony overgrowth may be observed on the leg bones, jaw bones and ribs. Their appearance indicates that there has been a high fluoride intake over a long period.

Lameness may occur as a result of this overgrowth on the leg bones. This may be followed by stiffness of joints. Effects of fluorides on skeletal structures are usually of a permanent nature.

Symptoms of advanced fluorosis include lack of appetite, general ill-health due to malnutrition, lowered fertility, reduced milk production and growth retardation. The symptoms developed in other species such as rabbits, horses and poultry are similar to those in cattle and sheep, i.e., mottling, staining and wearing of teeth, bony overgrowths on the skeleton, stiffness and lethargy and general ill-health from malnutrition and starvation.

While diagnosis of chronic fluorine poisoning is done, one has to be careful. This is because drinking water supplies may contain unduly high levels of fluorides. Also, mineral supplements may contribute appreciable amounts of fluorine to the diet. Hence, a check of all possible sources must be made when investigating fluorosis.

**Arsenic**

Arsenic occurs as an impurity in many ores and in coal. It has been reported to cause poisoning of livestock near various industrial processes and smelters. Just like most industrial air contaminants, arsenic may spread over a considerable area from a stack source. Arsenic in dusts or sprays on plants can lead to poisoning of cattle.

**Symptoms of Acute Arsenic Poisoning**

In acute cases the symptoms are severe salivation, thirst, vomiting, uneasiness, feeble and irregular pulse and respiration. The animal stoops, lies down and gets up. There is diarrhoea, and the faeces have a garlic odour and are sometimes bloody. The ears become cold, the body trembles and develops abnormal temperature and convulsion. Death may occur in few hours or days.

**Symptoms of Chronic Arsenic Poisoning**

Arsenic appears to have a depressing effect upon the central nervous systems. The animal becomes dull, and exhibits a lack of appetite, with a resulting weight loss. Also, there may be chronic cough and diarrhoea may occur continuously. There may be thickening of the skin, anaemia and abortion or sterlity. Chronic poisoning can result in eventual paralysis and death.

**Tolerance limit**

Arsenic as well as its soluble compounds are extremely poisonous. It has been reported that sheep have been poisoned by as little as 0.25-0.50 g of arsenic daily, whereas cattle and horses may tolerate 1.3-1.9 g daily. It has also been reported from laboratory experiments, that 10 mg of arsenic per kilogram of body weight per day has produced chronic arsenic poisoning in guinea pigs. A detailed investigation of the effects of arsenic poisoning on farm animals is necessary to establish standard tolerance limits.

**Lead**

Lead contamination of the atmosphere takes place on account of various industrial sources such as smelters, coke ovens and other coal combustion processes. Lead is also used in dust and sprays containing lead arsenate

**Symptoms of Acute Lead Poisoning**

In case of acute lead poisoning the onset is sudden and the course relatively short. Prostration, staggering and inability to rise are prominent symptoms. The pulse is always fast but weak. Some animals may fall suddenly, stiffen the legs and have convulsions. There is complete loss of appetite, paralysis of the digestive tract and diarrhoea. Other nervous symptoms in cattle are grinding of the teeth and rapid chewing of the cud.

In case of horses, it may result in complete loss of appetite, nervous depression, lethargy and death.

**Symptoms of Chronic Lead Poisoning**

Chronic lead poisoning has been observed frequently in horses that have been grazing on forage near smelters, lead mines, and in orchards that have been sprayed. Paralysis of the muscles of the larynx and difficulty in breathing are the main symptoms. Convulsions may occur-from the paralysis of the throat, and the difficulty in breathing may be unusually severe and persistent, during and after exercise.

**Tolerance Limits**

Lead is a cumulative poison. Therefore, the continuous ingestion of even very small daily doses will be finally as effective as one toxic dose. Hence in case of slight contamination death can occur after many months and in case of severe contamination death can occur within 24 hours.

It has also been reported that inadequate calcium in diet increases the retention of lead in the animal body. Low—calcium diets may result in lead storage of as much as five times as compared to that found in animals that receive sufficient amounts of calcium. However, extra quantities of calcium above the amount considered sufficient, will not offer additional protection against poisoning.

**Effects of Air Pollution on Plants**

Air pollution has been known to have an adverse effect on plants. At first, it was only sulphur dioxide that was considered a dangerous pollutant. Now, with the advent of various pesticides and new industrial processes, the range of harmful pollutants has multiplied tremendously. Sometimes, vegetation over 150 km away from the source of the pollutant has been found to be affected.

Industrial pollution, particularly from smelters, has caused complete destruction of vegetation in some cases e.g., at Ducktown, Tennessee. The international dispute between Canada and the United States over the damage caused by the Trail, British Columbia Copper Smelter is well known. Los Angeles smog has caused widespread damage to some crops and forests in Southern California. In fact, many books and reports have been published in USA on air pollution injury to vegetation. In some cases, economics of the damage caused to crops and plants has been estimated. Compensation amounting to thousands of dollars has been paid for damage done to the crops. In our own country, there are many reports of the effect of pollutants like cement dust on plants.

**Effect of Environment on Plants**

The primary factor which controls gas absorption by the leaves is the degree of opening of the stomata. When the stomata are wide open, absorption is maximum and vice versa. Consequently, the same conditions that enhance the absorption of the gas ($CO_2$ for photosynthesis), predispose the plant to injury (by absorbing a pollutant gas like $SO_2$). The conditions that cause the stomata to open are high light intensity (especially in the morning hours), high relative humidity, adequate moisture supply to the roots of the plant and moderate temperatures.

Most plants close their stomata at night and are therefore much more resistant at night than in the day time. But some plants like the potato, which do not close their stomata at night are as sensitive in the dark as in the light.

**Air Pollutants Affecting Plants**

| | |
|---|---|
| Sulphur dioxide | Fluoride compounds (like hydrogen fluoride) |
| Ozone | Chlorine |
| Hydrogen chloride | Nitrogen oxides (NO, $NO_2$, etc.) |
| Ammonia | Hydrogen sulphide |
| Hydrogen cyanide | Mercury |
| Ethylene | PAN (peroxy acetyl nitrate) |
| Herbicides (sprays of weed killers) | Smog |

The above pollutants interfere with plant growth and the phenomenon of photosynthesis. Smog, dust, etc., reduce the amount of light reaching the leaf and also by clogging the stomata may reduce carbon dioxide intake to some extent and thus interfere with photosynthesis.

**Forms of Damage to Leaves**

Damage to leaves takes several forms

| | |
|---|---|
| 1. Necrosis | Necrosis is the killing or collapse of tissue |
| 2. Chlorosis | Chlorosis is the loss or reduction of the green plant pigment, chlorophyll.<br>The loss of chlorophyll |

| | | |
|---|---|---|
| | | usually results in a pale green or yellow pattern. Chlorosis generally indicates a deficiency of some nutrient required by the plant. In many respects, it is analogous to anaemia in animals. |
| 3. | Abscission | Leaf abscission is dropping of leaves. |
| 4. | Epinasty | Leaf epinasty is a downward curvature of the leaf due to higher rate of growth on the upper surface. |

Tissue severely injured by air pollutants often has a characteristic colour. Bleaching is associated with sulphur dioxide, yellowing with ammonia, browning with fluroide. A silvering or bronzing of the under surface of some leaves is associated with PAN injury. Phytotoxicant is the name given to plant damaging substances.

### Kinds of Injury to Plants

#### *Acute Injury*

It results from short-time exposure to relatively high concentrations, such as it might occur under fumigation conditions. The effects are noted within a few hours to a few days and may result in visible markings on the leaves due to a collapse and death of cells. This leads to necrotic patterns, i.e., areas of dead tissue.

#### Chronic Injury

It results from long-term low level exposure and usually causes chlorosis or leaf abscission.

**Growth or Yield Retardation**

Here the injury is in the form of an effect on growth without visible markings (invisible injury). Usually a suppression of growth or yield occurs.

While deciding about the type of injury the plant has suffered, one must carefully distinguish between adverse effects caused by air pollution and adverse effects resulting from other factors. Plant diseases may show symptoms which are very similar to those caused by air pollution. High temperatures, poor plant care, shortage of nutrients and water may also cause an appearance which is similar to that of a plant damaged by air pollution. High temperatures, poor plant care, shortage of nutrients and water may also cause an appearance which is similar to that of a plant damaged by air pollution. It may also be due to insect damage. Therefore, while diagnosing the effects of air pollution one must consider factors such as plant disease, nutrition history, weather damage, insect damage, and the nature of pollutants in the area.

**Effects of Air Pollutants on Plants**

The effects of various pollutants like sulphur dioxide, ozone, fluorides, etc., on plants is shown in Table.3

**Table 3:** *Effects of Pollutants on Plants*

| | Pollutant | Dose | Effect |
|---|---|---|---|
| 1. | Sulphur dioxide | Mild | Interveinal chlorotic bleaching of leaves Severe Necrosis in interveinal areas and skeletonized leaves |
| 2. | Ozone | Mild | Flecks on upper surfaces, premature aging and |

| | | | |
|---|---|---|---|
| | | | suppressed growth Severe Collapse of leaf necrosis and bleaching |
| 3. | Fluorides | Cumulative | Necrosis at leaf tip effect |
| 4. | Nitrogen dioxide | Mild | Suppressed growth, leaf bleaching |
| 5. | Ethylene | Mild | Epinasty, leaf abscission |
| 6. | PAN | Mild | Bronzing of lower leaf surface (upper surface normal), suppressed growth, Young leaves more susceptible |

**Effect on Sulphur Dioxide**

Sulphur dioxide produces two types of injury on the leaves of plants- acute and chronic, depending on the concentration and period of exposure. The acute injury is characterized by the killing of marginal or interveinal areas of the leaf. Immediately after fumigation these areas will get a dull, water - soaked appearance, subsequently they dry up and usually bleach to an ivory colour, though some species finally assume a brown or reddish-brown colour. Chronic injury is caused by the slow, long -continued absorption of sub-lethal amounts of gas or by absorption of an amount of gas somewhat less than that necessary to cause acute injury.

Sulphur dioxide is phytotoxic in concentrations above 0.1-0.2 ppm. Below about 0.4 ppm. it tends to be oxidised in the cells as rapidly as it is absorbed, but interferences with functions such as photosynthesis is slight. Chronic injury, if any, is generally exhibited with these small concentrations.

Above about 0.4 ppm, acute injury occurs more frequently. Sometimes temporary interference with photosynthesis or 'invisible injury' can occur. However, the mechanism by which sulphur dioxide affects the plants has not yet been well understood, though various explanations have been put forth by many investigators.

**Effect of Hydrogen Fluoride**

Hydrogen fluoride behaves somewhat similar to sulphur dioxide except that with a few species of plants it is effective in causing lesions and interfering with photosynthesis in concentrations two or three orders less than in the case of sulphur dioxide. With most species it is up to 10 times as effective as sulphur dioxide. However, recovery of plants from the fluoride effect is much slower than from sulphur dioxide. This difference in rate of recovery is probably explained by the fact that sulphite in the leaves is rapidly oxidised to the relatively non-toxic sulphate, whereas fluorides can be removed only by the slower process of volatization or by some obscure chemical reaction. Forage may be rendered unsafe for animal feeding if more than 50 ppm of fluorine is absorbed.

**Effect of Ozone**

Ozone is phytotoxic in exposures of a few hours at about 0.2 ppm. Injury due to ozone is quite different from typical smog injury. The lesions are generally confined to the upper surface. The effects of ozone on plants in mild and severe doses is indicated in Table 3.

**Economic Effects of Air Pollution**

Air pollution damage to property is very important economic aspect of pollution. In the United States of America, Great Britain, France and many other

European countries, this problem has been investigated in detail and successful attempts have been made to translate observable air pollution damage into terms of economic impact. Air pollution damage to property covers a wide range - corrosion of metals, soiling and eroding of building surfaces, fading of dyed materials, rubber cracking, spoiling or destruction of vegetation, effects on animals, as well as interference with production and services. Another important economic effect of air pollution is deterioration of works of art. In our country today there is an urgent need to investigate and study this problem in detail and express the damage to property in economic terms, as very little work has been done in this direction so far.

**Mechanisms of Deterioration in Polluted Atmospheres**

Air pollutants cause damage to materials by five mechanisms:

1. *Abrasion*: Solid particles of sufficient size and travelling at high velocities can cause abrasive action. Also, large sharp edged particles embedded in fabrics can accelerate wear.
2. *Deposition and removal:* Solid and liquid particles deposited on a surface may not damage the material itself but it may spoil its appearance. However, the removal of these particles may cause some deterioration. Although a single washing or cleaning may not cause noticeable deterioration, frequent cleaning ultimately does.
3. *Direct chemical attack:* Some air pollutants react directly and irreversibly with materials to cause deterioration. For example, the bleaching of marble by sulphur dioxide, the tarnishing of silver by hydrogen sulphide and the etching of a metallic surface by an acid mist.

4. *Indirect Chemical attack:* Certain materials absorb some pollutants and get damaged when the pollutants undergo chemical changes. For example, sulphur dioxide absorbed by leather is converted to sulphuric acid, which deteriorates the leather.

5. *Corrosion:* the atmospheric deterioration of ferrous metals is by an electrochemical process, i.e corrosion. This due to the action of air pollutants facilitated by the presence of moisture.

Table 4 presents a generalized picture of air pollution damage to various materials.

**Table 4:** *Air Pollution Damage to Various Materials*

| Materials | Principal air pollutants | Effects |
|---|---|---|
| Metals | $SO_2$ acid gases | Corrosion, spoilage of surface, loss of metal, tarnishing |
| Building materials | $SO_2$ acid gases, | Discolouration, leaching |
| Paint | $SO_2$, $H_2S$ particulates | Discolouration |
| Textiles and | $SO_2$ acid gases and | Deterioration, reduced tensile strength, |
| Textile dyes | $NO_2$ Ozone | and fading |
| Rubber | Oxidants, ozone | Cracking, weakening |
| Leather | $SO_2$, acid gases | Disintegration, powde-red surface |
| Paper | $SO_2$, acid gases | Embrittlement |
| Ceramics | Acid gases | Change in surface appearance |

**Factors Influencing Atmospheric Deterioration**

There are four prime factors which influence the rate of attack of damaging pollutants. They are moisture, temperature, sunlight, and air movement.

1. *Moisture:* The presence of moisture in the atmosphere greatly helps the process of corrosion. Without moisture in the atmosphere, there would be little atmospheric corrosion, if any, even in the most severely polluted environments. In case of sulphur dioxide and various combinations of particulates, investigations have revealed that the rate of corrosion of metals will increase as relative humidity in the air increases.

2. *Temperature:* Temperature affects the rate of the chemical reaction and consequently it affects the rate of deterioration.

3. *Sunlight:* In addition to producing damaging agents such as ozone through a series of complex photochemical reactions in the atmosphere, sunlight can cause direct deterioration of certain materials. But in cases of fading of certain dyes or rubber cracking, damage due to direct sunlight cannot always be distinguished from that caused by ozone.

4. *Air movement:* Wind direction is an important factor to be considered in places where deterioration is caused by pollutants released from nearby factories. For example, damage to agricultural crops in the vicinity of factories. Similarly, wind speed is also an important factor in determining the impact of air pollutants on the receiving surfaces. Pollutants present in wind blowing at high speeds will have more abrasive effects and they may also be carried over long distances. Where leather bound books are stored,

air movement is a critical factor when the air is contaminated by sulphur dioxide.

**Effects on Art Treasures**

Air pollutants have caused great damage to art objects throughout the industrialized world. Some excellent monuments of antiquity have been silently eaten away by pollutants. This can be clearly seen in many works of art sculptures and carvings on historic buildings and cathedrals. A few examples of such works of art which have been affected by air pollution both in India and abroad have been described below.

**Effect on Art Treasures in India**

(a) *Taj Mahal:* The Taj Mahal a miracle and marble and a historical movement is facing grave danger from pollution created by existing foundries, power house, railways yards and other industrial units. According to some environmentalists, the problem now seems to have been aggravated because of the commissioning of the Mathura Oil Refinery, about 40km from the priceless monument. Whether the monument is going to be affected by the pollutants emitted from the refinery or not has created a controversy which is now being debated. Many reports on this have appeared in noted newspapers and journals. However, it is to be hoped that certain anti-pollution safeguards built into the refinery will minimise the effect of pollutants on the monument.

According to some reports, pollutants from existing foundries power houses and railway yards are already affecting the Taj Mahal. Black and brown patches are disfiguring the shiny white surface of the Taj. Reports in newspapers also speak of the formation of gypsum, scaling and flaking effects.

Sulphur dioxide from the various sources mentioned, mixes with the atmospheric moisture and gets converted into sulphuric acid which settles down on the exterior of the Taj Mahal. The acid reacts with the marble (calcium carbonated) and corrosion and discolouration of the monument is the ultimate result.

It looks as though more money will have to be spent on protecting the Taj Mahal from the effects of pollution than it took Shah Jehan to construct it. Perhaps the immediate solution to save the Taj Mahal lies in shifting the foundries and power houses from Agra, and in effectively supervising anti-pollution measures adopted by the refinery. Further, the coal locomotives being used by the railways in that region can be replaced by diesel engines. In addition to these steps, a dense green belt of suitable tree species should be created around the Taj Mahal to absorb atmospheric pollutants as trees can act as effective air scrubbers.

(b) *Belur temple:* The renowned temple of Sri Channakeshava at Belur (Hassan district, Karnataka state) is threatened with a similar hazard. A plywood factory located close to the temple emits soot-laden fumes which get deposited on the sculptures in the temple and discolour the surface, inside and outside. This pollution effect is under investigation.

**Effect on Art Preasures in Other Countries**

The Parthenon in Athens, has been affected in recent years because of air pollution, Cleopatra's needle the large stone obelisk moved from Alexandria, Egypt, to London, has suffered more deterioration in the damp, smoky, acid atmosphere of London in the last 90 years than in the eariler 3000 or more years of its history. Another ancient Egyptian obelisk in New York City, shown similar decay. Similar cases have been reported from other parts of the world. The colosseum in Rome, and the San Marco Basilica in Venice, Italy, show signs of decay, possibly due to air pollution, In France, it is reported that environmentalists have removed statues from the exterior of cathedrals and replaced them with copies. Similarly ancient temples and buildings in industrial areas of Japan are facing the hazards of atmospheric pollution. Paintings, books fabrics, antique costumes and other art objects have suffered damage.

Of course, it is impossible to make adequate substitution or compensation for the losses suffered by works of art. They are irreversible losses. It is impossible to estimate the losses in economic terms for these intangible assets. In some cases it is possible to preserve them, by applying speicial preservatives. But they are not always successful. Sensitive art objects displayed inside buildings can be placed in hermetically sealed containers. Air conditioning can also be used as a protective measure.

# 2

# Water Pollution

Water pollution is a phenomenon that is characterized by the deterioration of the quality of land water (rivers, lakes, marshes and ground water) or sea water as a result of various human activities. Here, we will first classify various human activities that cause water pollution, discuss water pollution according to their characteristics, and then take a general view of water pollution which is currently in progress in our modern society. Human activities related with water pollution comprise various industries such as mining, agriculture, stockbreeding, fisheries, forestry, urban human activities, manufacturing industry, construction works and various tertiary industries. Then, in the last section, we will classify waste waters according to their physical and chemical properties.

Water pollution is any physical or chemical change in water that can adversely affect organisms. It is a global problem, affecting both the industrialized and the developing nations. The water pollution problems in the rich and the poor nations, however, are quite different in many respects. Heat, toxic metals, acids, sediment, animal and human wastes, and synthetic organic compounds foul the waterways of developed nations. Human and animal wastes, sediment, and pathogenic organisms head the list in the nonindustrialized nations. In these countries, unsanitary water and malnutrition account for most of the illness and death.

Water has many different uses, each requiring different levels of purity. For example, water from the Ohio River may be clean enough to wash steel but may be toxic to fish and wildlife: water suitable for boating or fishing may be unsuitable for swimming : water safe to swim in may be too polluted to drink.

Like air pollutants, water pollutants come from numerous natural and anthropogenic sources. Likewise, water pollutants produced in one nation may flow into others, creating complex international control problems that may take decades to solve.

**Major water pollutants**

Following are the eight major types of water pollutants:

1. Oxygen-demanding wastes (domestic sewage, animal manure, and some industrial wastes)
2. Disease-causing agents (bacteria, parasites, and viruses)
3. Inorganic chemicals and minerals (acids, salts, and toxic metals)
4. Organic chemicals (pesticides, plastics, detergents, industrial wastes, and oil)
5. Plant nutrients (nitrates and phosphates)
6. Sediments (soils, silt, and other solids from land erosion)
7. Radioactive substances
8. Heat (from industrial and power plant cooling water)

Although sedimentation is the major water pollutant in terms of quantity, other water pollutants can pose serious threats to plants and animals (including humans) when present in minute quantities. Some organisms have low tolerance levels to chemicals such as oil, some pesticides, and lead and mercury compounds. Other organisms with higher tolerance levels can be affected when the levels of some pollutants are biologically concentrated in food chains.

Water pollutants can be classified as degradable (either rapidly or slowly), or nondegradable *Rapidly degradable* (non persistent) pollutants can be broken down fairly quickly by natural chemical cycling processes as long as the pollutants do not overload the system. They include domestic sewage, other oxygen-demanding wastes, plant nutrients, and some synthetic organic chemicals. To control these pollutants we need to prevent overload. Normally these pollutants are degraded into harmless or less harmful forms, but sometimes they are converted into more harmful forms. For example, microorganisms in acidic waters can convert moderately harmful mercury metal and inorganic mercury compounds into very toxic organic methyl mercury.

*Slowly degradable* (persistent) *pollutants* remain for long periods of time but are eventually broken down or reduced to harmless levels by natural processes. Examples include some radioisotopes and many synthetic organic chemicals, such as DDT, PCBs, phenols, and older-type detergents. Some of these substances, such as DDT, may be magnified up the food chain to levels that are harmful for wildlife and possibly for humans because most of these pollutants are extremely difficult to remove by waste treatment methods, control measures must emphasize (1) preventing or minimizing their entry into the environment, (2) long-term storage until levels are safe, and (3) research to determine which of these pollutants are dangerous and capable of being magnified up the food chain. *Nondegradable pollutants* are not broken down by natural purifying processes. They include some metals (such as mercury, lead, and arsenic), some salts of metals, sediments, some synthetic organic compounds (such as plastics), and some bacteria and viruses. These pollutants must be controlled either by removing them through waste treatment or by preventing them from entering the environment.

**Sources of water pollution**

Water pollution can result from natural runoff; dissolved chemicals in water that percolates through the soil; and human sources, such as agriculture, mining, construction, industry, homes, and businesses. It is useful to classify pollution sources as either point sources or nonpoint sources. Point sources involve discharge or wastes from identifiable points. They include (1) sewage treatment plants (which remove some but not all pollutants) (2) storm water runoff from combined storm and sanitary sewer lines in urban areas; (3) industrial plants; and (4) animal feedlots where large numbers (sometimes 10,000 or more) of animals are scientifically fed in tightly restricted quarters before slaughter. Nonpoint sources involve the diffuse discharge of wastes from land runoff, atmospheric washout, and sources that are difficult to identify and control. They include (1) runoff of sediment from natural and human caused forest fires, construction, logging, and farming; (2) runoff of chemical fertilizers, pesticides, and saline irrigation water from croplands; (3) urban storm water runoff; (4) drainage of acids, minerals, and sediments from active and abandoned mines; and (5) spills of oil and other hazardous materials. Nonpoint source pollution is not recognized as a major problem, since the sources are widely spread out, difficult to identify, and hard to control.

**Water pollution by agriculture**

Agriculture has been a victim of water pollution in many instances, but sometimes, it is also responsible for polluting water. Water pollution caused by agriculture is mainly an outcome of fertilizers and agricultural chemicals such as insecticides and herbicides.

***(1) Water pollution by fertilizers***

Fertilizers given to crops are not always fully consumed, and part of it remains in the soil by being absorbed by soil colloid, and influence the quality of underground, river and sea waters when it is dissolved. Chemical fertilizers mainly consists of relatively simple compounds of nitrogen, phosphorus and potassium which are the nutritive elements of plants. Their runoff coefficients vary depending on the solubility of the fertilizer itself, the rate of absorption by plants, rate of absorption by the soil and rate of decomposition. It is generally believed that the runoff coefficient of nitrogen is 5-30%, and that of phosphorus is 0.5-5%. Especially, nitrogen having a high runoff coefficient tends to cause the eutrophication of lakes and enclosed sea areas.

Fertilizers as the sources of water pollution are called "non-point sources" in contrast to "point sources" such as sewage and industrial waste waters. The non-point sources are not only hard to eliminate, but also hard to measure their quantity. Organic fertilizers such as compost, farmyard manure, human waste and fish meal produce nitrogen and phosphorus when they are decomposed, and also become the sources of pollution. These organic fertilizers do not contain, in general, any toxic substances.

***(2) Water pollution by agricultural chemicals***

Agricultural chemicals are considered as a source of peculiar pollution which cannot be overlooked. Agricultural chemicals comprise a variety of chemicals such as insecticides, sterilizers and weed killers, and have played important roles in the operation of modern agriculture, but they have also caused serious environmental pollution.

The characteristics of agricultural chemicals as water-polluting substances lie in that (1) almost all of them are special compounds not occurring at all in nature, and many of the behaviours and changes of such chemicals at the time when they are discharged into the natural environment are known; (2) many of their influences on the human body because of their inherent characteristic; (3) when they are ingested into the human body, they tend to accumulate or concentrate in it; (4) they are available in a variety of kinds, and new products are developed constantly; (5) some of them are extremely stable as chemicals, and remain in nature or in the human body for a long time thereby eventually afflicting the human body; (6) it is common to all such chemicals to contain additives besides the main constituents, and the additives can contribute to affect the environment though they do not have a strong toxicity; (7) many of them require a special method of analysis.

Of the agricultural chemicals developed in the past, many of them are prohibited from being used because of their violent toxicity. Nevertheless, those remaining in the environment are still factors that contribute to environmental pollution.

In addition to above-mentioned pollutions, there are instances of water pollution caused by indirect effect of agricultural activities. For example, the change in the condition of the ground surface by reclamation of waste land, deforestation, supply of the soil to low land, and the alteration of the natural water system for irrigation purposes can be considered indirect causes of water pollution.

**Water pollution by mining industry**

In many countries, water pollution related to the mining industry can be traced back to olden times compared with those caused by other industries. The characteristics of waste water from mining industries varies greatly depending on the kind of mine. For instance.

***(i) Metal mining industry***

Many kinds of waste waters are discharged. For example, mine water, ore dressing water, waste water from smelters, etc. In general, it is strongly acidic (low pH), and contains metal irons, sulphuric acid and suspended solids of high concentrations.

***(ii) Coal mining industry***

Waste water from coal mines mainly consists of mine water, coal dressing water, exudative water from coal-waste heap etc.. and contains a large quantity of suspended solids. In general, the scales of operation of coal mines is much larger than that of other mining industries, and thus the coal mining industry exerts a large influence upon the natural environment.

***(iii) Sulphur mine***

In general, waste water from sulphur mines contains large quantities of sulphuric acid and iron and has a low pH value. Almost all sulphur mines are deserted; so the administration of deserted mines becomes a social and environmental problem.

***(iv) Petroleum and natural gas drilling industries***

Different from other mining industries, these industries are very often located in plains. Waste water generally consists of welled-up mine water and waste water incidental to the drilling operations. It contains large quantities of salt and organic matters such as mineral oil.

***(v) Stone-quarrying and clay-supply industries***

These industries supply lime stones, building stones, porcelain clay etc. Waste water from the sites of these industries generally contains large quantities of suspended solids.

The characteristics of industrial pollution common to the above-mentioned mining industries can be summarized as follows:

(i) In many instances, the development of a mine itself changes its natural environment artificially, and such a change first affects the quality of water of the environment even if any kind of waste water is not discharged from mining activities.

(ii) In areas where mines can be developed, very often, even before the development of the mine, the quality of river water or underground water is different from that of other areas. For instance, river water or underground water in such areas contain special metals of high concentrations and is strongly acid. Thus, it is rather difficult to measure the direct influence of mining activities on the quality of water. And also, geological features and the ways of land use are usually very peculiar in mining areas. These conditions should be taken into consideration prior to surveying the quality of water.

(iii) Mining activities generate large quantities of solid wastes in most cases, and such wastes from large heaps. Such wastes not only pollutes water by exudation of toxic substances, but also, the rapid generation of suspended solids in water when the heap is washed away by a heavy rainfall or a flood.

**Water pollution by stockbreeding and fisheries**

1. When cattle or horses are fed on a big pasture, there is little possibility of the livestock causing water pollution because their excreta are spread widely all over the pasture. However, when livestock is fed intensively in a limited area it causes a number of problems. In the case of the stockbreeding industry, the excreta and leftover feed on livestock are the sources of water pollution. Water pollution caused by stockbreeding has the following characteristics.

    (i) Degree of pollution varies largely depending on the location of the stockbreeding farm (whether in suburban or country area), system and scale of operation and the level of breeding technique.

    (ii) Composition and quantity of excreta of livestock varies depending on the type, quantity and methods of feeding and watering.

    (iii) Besides water pollution, the stockbreeding industry also generates an offensive odor, flies, noise and so on.

2. Besides stockbreeding farms, waste water from slaughterhouses is a source of water pollution which should not be overlooked.

3. In most instances, fisheries has been the victim of water pollution. But in recent years, as the number of fish farms increase, pollution of water areas surrounding fish farms caused by leftover feed and excreta of fish is becoming serious.

**Water pollution by urban activity**

In a sense, the existence of human beings itself is a source of water pollution. In primitive ages, the purifying capacity of nature was much greater than the rate of water pollution caused by human living. But as time passed, men began to settle near water areas for the convenience of living; then they developed various industries which provided them with the foundation of civilized life. The emergence of various industries coupled with the development of civilization has constantly increased the quantity of waste water containing pollutants. On the other hand, the increasing concentration of the population has also invited the concentration of pollution sources. As a result, the rate of environmental pollution has exceeded the rate of natural purification.

If all the urban activities of human beings related with water pollution are viewed as a metabolic system, it can be outlined as follows:

| | |
|---|---|
| Waste water | |
| Water | Human waste and municipal waste |
| Energies | Waste gas and waste heat |

Substances Municipal wastes such as waste water, human waste, household waste, waste gas and waste heat are the causes of water pollution.

The type of water pollution in a city varies largely depending on the characteristics of the city, life style of the inhabitants, degree of civilization, degree of development of sewage treatment plants or sewerage systems. Although, it is very difficult to investigate these conditions in detail, they can be estimated by investigating the following:

(i) Population of the city: This is the most essential datum to be obtained first.

(ii) Consumption of water and quantity of waste water: where a city water system is well established, it is relatively easy to obtain data concerning the consumption of water. For example, the consumption of city water in a modern civilized city is said to be about 500 liter/man/day. However, it is usually difficult to know the quantity of waste water even if it is only an estimate.

(iii) Condition of sewage treatment: This can be measured in terms of the availability of sewage treatment plants and sewage works including their capacities and efficiency.

(iv) Condition of municipal or public facilities and private facilities: In a large city, there are a number of public facilities and various other facilities, and all these facilities can be the sources of water pollution, such as: (public facilities) Government and municipal offices, schools, hospitals, public bath houses, public lavatories, markets, ports, power plants, city gas facilities, sewage treatment plants. etc.

(Non-public facilities) Factories, stores, offices, gas stations, restaurants, bars, hotels, etc.

(v) Condition of sewage: Municipal sewage mainly consists of a· mixture of human waste, household sewage and wastes from other facilities, and its condition varies depending on the characteristics of the city, degree of civilization, precipitation and the condition of road surface.

The quantity and condition of waste water resulting from human activity, mainly consisting of household waste water, generally vary depending on the time in a day.; however, the "product of the concentration multiplied by the quantity" tends to remain within a relatively fixed range. Such a value is defined as "pollution loading amount", and is used as the basic data for the sewage treatment planning.

**Table 1: Standard values of BOD, SS and pollutant load of sewage and human waste**

| *Type of waste water* | *Load, equivalent, concentration, etc. of BOD and SS* |
|---|---|
| Human waste from flush toilets (including those from septic tanks for sewage) | (1) BOD equivalent per person per day: 13 g (BOD: 260 mg/) |
| | (2) SS equivalent per person per day: 26 g (SS: 520 mg/1) |
| | (3) Flushing water per person per day: 50/ |
| Household waste water (not Including rainwater and special waste water) | (1) BOD equivalent per person per day: 40 g (BOD: 200 mg/1) |
| | (2) SS equivalent per person per day: 50 g (SS: 250 mg/1) |
| | (3) Quantity of waste water per person per day: 200/ |
| Other typical waste waters | (1) Average BOD of waste water from office building; about 80 mg/1 |

(2) Average BOD of waste water from hotels about 100 mg/1
(3) Average BOD of waste water from hospitals about 100 mg/1
(4) Average BOD of waste water from restaurants and catering facilities: about 400 mg/1

---

The most conspicuous characteristics of municipal sewage are that the organic content (indicates as BOD) is extremely high; the contents of ammoniacal and organic nitrogen, phosphates and sulfides are considerably high; and normally, hazardous substances are not contained

Industrial waste water tends to contain hazardous substances; therefore, its discharge into the municipal sewerage system should be strictly controlled. This is necessary because ordinary sewage treatment plants not only do not have an equipment to remove hazardous substances, but also there is a possibility of getting its activated sludge processing function damaged by such hazardous substances.

The modern sewage treatment technique is capable of removing organic pollutants very efficiently, but it is difficult economically to re-move nitrogen and phosphorus contents efficiently with existing facilities. For example, only about 30% of the nitrogen and phosphorus contents can be removed through the currently employed two-step treatment method, which mainly applies the activated sludge method, and the remainder of such contents causes the eutrophication of water areas.

Anyhow, all of the above mentioned techniques are possible where the municipal sewerage system and sewage treatment plants are well established, and the quantity and quality of waste water can be measured effectively. However, where sewage treatment plants are not available, or where waste water is discharged directly into the hydrosphere without passing a sewage treatment plant, it is almost impossible to measure the quality and quantity of waste water.

In case waste water is directly discharged into the hydrosphere, and if the pathogenic bacteria of an infectious disease is contained in the human waste, such bacteria would be discharged into the hydrosphere in active condition thereby eventually spreading the infections disease through the water system. The degree of such a danger various depending on the quality of sanitation facilities of individual cities.

**Water pollution by manufacturing industry**

Characteristics of water pollution by Manufacturing Industry

Of water pollutions with which human activities are concerned, the manufacturing industry is the largest in scale, and most diversified. The characteristics of the water pollution of manufacturing industry can be summarized as follows:

1. Unlike the mining industry and agriculture, the discharging condition of waste water from manufacturing industries is relatively simple because it is rarely affected by complicated conditions of the natural environment. For example, waste water of the manufacturing industry is discharged concentratively from a fixed outlet, which is called "point source pollution". Besides, waste water from the manufacturing industry is generated as a result of the production activity which is 100% artificial, and thus the manufacturing industry can be considered as the source of water pollution which is most easy to catch and control in this point.

2. Sources of water pollution from manufacturing industry are relatively easy trace back through scientific method. But their control is not always easy because it involves complex social and political problems. Furthermore, tracing such point sources of pollution is not always easy because the interests of individual manufacturing companies can be affected by pollution control measures.

3. In contrast to the relatively simple form of discharging process, the composition and quantity of waste water vary widely depending on the type and scale of individual factories, change from time to time, and its influence also takes a variety of forms.

4. The influence of waste water on the environment varies depending on the geographical location of the factory. For example

   (i) In the case of an industrial conglomerate or large industrial complex: The influence of waste water on the environment is extremely large, but as far as such manufacturing facilities are concentrated in one area based on the results of careful environmental assessment, waste water from these facilities can be controlled in the most ideal form;

   (ii) In case of factories are located in commercial or residential areas of a city: The location of this type is primarily undesirable. Actually,however, there are a number of such cases because of historical reasons, or geographical advantages concerning the purchase of raw materials, transportation of products, or access to the labour market. In Japan, locations within easy access to the source of industrial water are decreasing in these areas. As a result, factories that recycle water are increasing; small scale manufacturing companies establish cooperative associations and are promoting the construction of joint waste water treatment facilities;

(iii) Location of the manufacturing factory is whether mountainous area, plain or coastal region is also an important factor to be considered in assessing the influence of waste water on the environment. Especially, in the case of a water-consuming type of manufacturing industry, its location has to be decided depending on accessibility to an industrial water source. On the other hand, the condition of the river or the sea into which waste water of the factory has to be discharged also has a great influence on the form and type of environmental pollution.

**Characteristics of waste water of main industries**

As discussed in the forgoing, the composition of industrial waste water is complex, and it varies largely according to the difference in the manufacturing process even among the same type of manufacturing factories. Furthermore, the manufacturing industry is constantly subject to the impact to technological innovations which occur at increasingly short intervals. For example, in Japan, the technique for recycling of waste water is making rapid progress. Manufacturing companies were compelled to do so due to the increasingly strict pollution control requirements and the increasing shortage of industrial water. As a result, manufacturing companies are also compelled to alter conventional manufacturing processes. It is rather difficult to enumerate the characteristics of waste water of each industry General characteristics of waste water according to the type of manufacturing industry are as follows:

***(1) Food processing industry***

A common characteristic of waste water from all types of food processing industries is the high concentration of organic content. Another common characteristic is the high concentration of solid, semisolid and suspended solid contents in waste water discharged from raw material treatment processes. The quality and quantity of waste water vary depending largely on the season mainly due to the variation of raw materials to be processed. Materials contained in waste water can be divided roughly into carbohydrate, protein, fat and a mixture of them. In many instances, waste water contains oil, nitrogen and phosphorous. In the case of starch producing, super refining and brewing industries, the BOD level of waste water is especially high.

***(2) Textile industry***

There is much difference in the content of waste water between natural and chemical textile industries.Waste water from the wool industry contains high concentrations of BOD, fat and alkali. Waste water from dyeing discharges various pollutants such as dyes, auxilliary additives and other chemicals.

***(3) Paper and pulp making industries***

These are typical water-consuming industries and the levels of COD and suspended solids in waste water are extremely high. Especially, waste water from the digesting process contains all components of wood except cellulose and has high COD and dense colour, so that it is necessary to collect and treat most of the content to recover. A large quantity of waste paper fibre is also discharged as supended solids together with the waste water.

***(4) Petroleum refining industry***

The main component of pollutants is spilled oil; but waste water also contains hydrogen sulfide, other sulfides, ammonia, mercaptan, and phenols,etc.

***(5) Petrochemical industry***

The waste water contains hydrocarbons, various organic compounds and catalysts, and its content varies largely depending on the type of product.

***(6) Iron and steel making industry***

This industry is also a water-consuming industry. Waste water from the cooling and cleaning processes for coke furnace gas contains ammonia, cyanide, phenols, etc. Besides, there is waste water from the dust collecting process of each furnace which contains suspended solids (coke, dust and ore), and that from the pickling process which contains acid, iron and oil.

***(7) Electroplating industry***

Many of the raw materials and chemicals used in this industry contain toxic substances, and thus the waste water contains various heavy metals (cadmium, zinc, copper etc.), cyanide, hexavalent chromium, acids and alkalis.

***(8) Leather industry***

The levels of BOD is high in waste water which comes from tannin used for tanning and raws. It also contains chromium used for tanning and lime for pretreatment, suspended solids and dyes.

***(9) Steam and nuclear power plants***

These power plants discharge large quantities of

effluent water from the cooling process. In Japan, all the large power plants are located close to the sea, and these plants are using sea water for cooling. The temperature of waste cooling water is normally high, and such waste water can cause thermal pollution or the elevation of temperature to the vicinity of discharing outlet. No other pollution is considered.

**Water pollution by other human activities**

Aside from the various human activities already discussed in previous sections, there are other types of human activities which also cause water pollution, and such activities sometimes become social problems as sources of environmental pollution. These types of human activities are as follows:

***(1) Construction work***

Construction works, especially those large-scale land development projects or forestry conservation and flood control projects aimed at preventing natural disasters, sometimes can cause water pollution: Water pollution caused by such construction projects can be divided into those occurring incidental to the progress of construction works and those occurring due to the change in natural conditions as a result of the construction works. The latter is more important than the former. Construction works for a dam, river improvement (including those for the alteration of basins), port facilities, creation and reclamation of land and gravel gathering from river-bed are enumerated as examples of the above-mentioned causes.

***(2) Various tertiary industries***

Various tertiary industries can be considered to cause water pollution such as cleaning and laundry businesses, public bath houses, hotels, printing business, photographic film development businesses, hospitals, laboratories or research institutes, tourist businesses, transportation facilities and so forth. An increase in the load of municipal waste water as a result of the increase in the load of municipal waste water as a result of the increase in the working population of cities also has become one of the factors directly related to water pollution. Of various transportations, ships, especially tankers which discharge waste cleaning water, have come to the fore as sources of water pollution in coastal regions.

***(3) Errors, accidents and disasters***

Accidental damages to public sewage works or drainage of industrial effluent, as well as poor planning of sewage treatment plants, and erroneous operations of these facilities can also cause water pollution. Speaking of water pollution caused by accidents, pollution of the ocean due to oil resulting from the collision of tankers on the high seas has become an object of public attention in recent years.

**Classification of water pollutants by their physical and chemical properties**

By now we have taken a look at conditions of water pollution caused by human activities classifying by the type of activity. So as look at water pollution from a different angle, we may classify and explain it from the standpoint of its physical and chemical characteristics.

***(1) Suspended solids (Abbreviated as "SS")***

The causes what is called the "turbidity of water". In general, suspended solids are defined as those having a diameter of over 1u. Waste water from cement manufacturing industry and some of the mines can be enumerated as examples of waste water containing inorganic SS. Municipal sewage and waste water from kraft pulp factories are typical examples of those containing organic SS.

***(2) Organic matter***

Organic matter is usually measured as BOD or COD. Organic matter consumes the dissolved oxygen in water, and eventually destroys the ecosystem of the hydrosphere. This is the main content of not only municipal waste water, but also industrial waste water such as those from food processing, chemical, pulp and leather industries.

***(3) Inorganic matter***

Inorganic matter comprises acids, alkalis, various salts, metals, etc. Waste water containing such substances are those from mines, chemical, pulp, electroplating and other industries.

***(4) Inorganic and organic toxic substances***

Toxic substances are defined as chemical substances even traces of which adversely affect either the human body or the living environment.

***(a) Inorganic toxic substances***

Comprises metallic substances such as cadmium, lead, hexavalent chromium, mercury etc.., and non-metallic ones such as cyanide and arsenic; and are contained in various industrial and mining waste water.

***(b) Organic toxic substances***

The modern chemical industry has created a number of new synthetic substances and their special characteristics are utilized to develop new agricultural chemicals and medicines, etc. However, these synthetic substances adversely affect the environment when they are discharged into it. Many of these substances consists of organic compounds containing phosphorus, chlorine, mercury, etc.

***(5) Oil and fat (hexane Extractable Component)***

Comprise vegetable oils, animals bats and mineral oils. Oils and fats are found in municipal waste water and in waste water from food, oil and fat processing industries, etc. While mineral oils are found in the waste water from oil drilling, petroleum refining, petrochemical, iron and steel making and other industries. They are also found in waste water from ships. In Japanese standard method the content of oils and fats is measured quantitatively by extracting them with hexane and is called "hexane extractable component".

***(6) Substances contributing to eutrophication (Nitrogen Compounds and Phosphates)***

Salts such as nitrogen and phosphoric acid compounds $NH_4$,$NO_2$, and $PO_4$) exist in small quantities even in natural clean water, and these substances are ingested by organisms as nutrient salts; but they also contribute to the eutrophication of water areas when their quantities exceed certain levels these substances are discharged contained in municipal waste water, surplus of fertilizers, some industrial waste water and synthetic detergents.

***(7) Sulfur compounds***

Hydrogen sulfied ($H_2S$) is a typical sulfide that causes water pollution, and is formed in municipal waste water, human waste, waste water from the digesting process of pulp and waste water from gas plants. Hydrogen sulfide gives an offensive odor, and corrodes the facilities and buildings located near the passage of waste water because of its strong reducing power. The content of hydrogen sulfide indicates anaerobic index of the water area. In the volcanic zone, hydrogen sulfide frequently occurs naturally. Sulfur oxidés are main pollutants in air and fall to the earth as "acid rain".

***(8) Coloring and odor-giving substances***

Even if not directly harmful to human health, coloring or odor-giving substances in water areas can seriously influence the living environment of people. Such substances are found in the waste water from dyeing, chemical, agricultural product processing, marine product processing and other industries.

***(9) Coliform group***

Colon bacillus itself does not have any pathogenic nature, but it is used as an important index to measure pollution by human waste. Water pollution due to human excreta is caused mainly by the lack of a proper municipal sewerage system and is important since there is a possibility to contain pathogenic organisms.

***(10) Radioactive substances***

Waste water from research institutes or hospitals that use isotopes, nuclear fuel treatment plants and nuclear power plants are considered to include radioactive substances.

***(11) Warm and hot waste waters***

Warm or hot waste water consisting of use cooling water discharged from factories and steam or nuclear power plants can significantly influence the water area as part of the living environment when it is discharged in large quantities. However, when the temperature of such waste water is controlled properly it can be utilized for fish farming or district heating.

Table 5 shows the major sources activities that cause water pollution.

**Table 5: Major causes and sources of water pollution**

| *Cause* | *Source (industries and others)* |
|---|---|
| BOD Content | Food processing industry (canned goods, meat and dairy product (organic matter) processing, marine product processing. Starch process-ing), brewery, sugar making, wool processing (wool washing), pulp and paper industry, leather industry, municipal sewerage night soil treatment plant. Suspended solids (Organic SS) Food processing, paper and leather industries, coal (SS) washing, municipal sewerage. (Inorganic SS) Ceramic industry, gravel yard, quarrying industry, mining industry. Oils and fats (Oils, Food processing, petrochemi-cal, iron and steel making, metal fats, miniral Oils) finishing, wool processing (Wool washing) industries. Acids or Alkalis Mining, paper making, pulp, electroplating, metallurgical, chemical, textile and leather industries. Ammonia Gas and coke plants Phenols Gas and coke plants, synthetic resin and other organic chemical industries. |
| Sulphides | Pulp and gas industries, Sewage treatment plants, dyeing and leather industries. |
| Color | Dyestuff manufacturing, dyeing, paint manufacturing, leather, chemical industries, night soil treatment plants. |
| Odor | Chemical, pulp, marine product processing, meat processing, sugar making, leather industries, sewage and night soil treatment plants. |
| Detergent | (Surface Municipal sewerage, textile industry, laundry. active agent) |
| High-temperature | Steam and nuclear power water plants. |

**Eutrophication**

Phosphorus and nitrogen compounds (phosphates, ammonium salts, nitrites and nitrates, etc) are called nutrient salts which are essential to the growth of phytoplanktons and algae participating in the primary production. However, if they are excessively supplied together with organic substances, the production activities become too vigorous due to overnutrition, ultimately changing the aquatic ecosystem. This phenomenon is called eutrophication. Eutrophication of lakes is not an extraordinary phenomenon but one which occurs and slowly proceeds even in natural not polluted ecosystem. However, eutrophication in polluted sea and lakes progresses very quickly due to rapid increase of organic substances and nutrient salts, causing unusual growth of specific organisms and thus destroying the normal aquatic ecosystem. This type of eutrophication is often observed in closed water areas such as lakes and deeply recessed bays; the results is colouring and smelling of the drinking water taken from the lakes, lowered clarity of lake water, or the occurrence of the so-called "red tide" in sea and lakes as a visible and conspicuous phenomenon. The red tide is caused by unusual propagation of certain kinds of planktons at such a terrific speed as to change to colour of water.

Consequently, many environmental trouble occur such as death of fishes due to clogging of gills with planktons, oxygen deficiency over the polluted area, and generation of poisonous substances.

In Japan, this phenomenon has often been observed in the inland Sea and other closed water areas. The true causes have not yet been determined exactly, but it is generally accepted that phosphorus and nitrogen compounds contained in urban and industrial waste waters and playing an important role.

**Contamination of bottom sediments**

River, lake and sea bottoms are usually covered with a certain depth of muddy sediments. These sediments contain remains and excreta of aquatic organisms, other than the earth and sand carried over by rivers. Suspended solids in water tend to sediment remarkably in lakes, slow-flowing rivers, estuaries where the river stream is stagnated by tidal currents, and in closed sea areas.

Cases are also reported where some highly poisonous heavy metals and sparingly soluble organic substances, which had settled on the bottom of water and fixed as sediment over a long period, where dissolved again or raised under some conditions, causing serious hazards. A typical example is mercury in the bottom sediment of the Minamata Bay. It was about twenty years ago when mercury caused the Minamata disease, but mercury hazard are still continuing Minamata disease, but mercury hazards are still continuing even after the discharge of mercury from industrial waste water completely stopped.

At the bottoms of lakes and closed areas with high organic loads, dissolved oxygen is consumed rapidly during the putrefaction of organic substances contained in the bottom mud. If vertical mixing of water is insufficient due to stratification, oxygen dissolved on the surface of water from the atmospheric air cannot reach the bottom. In addition, poor clarity of water weakens sufficient penetration of the sun beam and significantly suppresses photosynthetic reactions in the bottom water layer. Under these conditions, the dissolved oxygen in the bottom water mass will decrease, ultimately leading to the zero-oxygen state. As a result, fishes in that area will be seriously, affected and the surrounding ecosystem may be materially destroyed.

Other causes are also known where elusion of phosphorus and nitrogen from bottom sediments in closed sea areas may be responsible for eutrophication and the resultant occurrence of red tide. Water bottoms often provide favourable spawning grounds for fishes; anaerobic conditions of bottom sediments destroy hatching in many causes.

A survey of bottom sediments provides the environmental information of the neighboring water area in the past over a considerably long period. Thus, it is possible to obtain new findings on the progress of pollution not expected from water quality investigations.

**Oil Pollution in oceans**

Modern civilization relies in a large measure on petroleum in all aspects. A huge amount of crude oil is carried by supertankers on oceans throughout the world, stored in tanks at terminal stations, and processed into various petroleum products at petrochemical complexes. Many terminal stations and petrochemical complexes are located in coastal zones. Probably a considerable amount of oil is always flowing into the ocean. Including slight leaks from various ground facilities, leakage from tankers during loading and unloading, and discharge of waste oil used on land. Besides, large scale oil spills due to maritime accidents have often occurred.

Oil spilt out to the sea tends to-spread on the sea surface as a thin layer. Volatile components gradually evaporate and some of them dissolve into water (benzene, for example, dissolves in seawater to a concentration of about 820 ppm at ordinary temperature). The oil membrane thus formed on the sea surface disturbs normal exchange of gases between the seawater and the atmospheric air, and also interrupts the penetration of sun beams to retard the growth of planktons.

The oil gradually becomes thick as the volatile components evaporate; the resultant viscous heavy oil is in or on the seawater, or gather shellfishes and sand on the sea bottom to form oil balls, which drift over long distance and coast shore. The spilt oil is degraded by some bacteria though very slowly. However, if it drifts faster than it decomposes, pollution will proceed. Heavy oil sometimes sticks to the bodies of seafowls and fishes; particularly, larvas are seriously damaged by oil deposition. Although not so critical, causes are often encountered when fishes cannot be used for food because of the smell of heavy oil.

# 3

# Noise Pollution

The planet, earth, is unique because the it is perhaps the only known human habitat so far. The man's environment is the basis of human existence and survival. The rapid technological developments no doubt have helped civilization but at the same time led to environmental degeneration. This in turn has threatened human survival. Noise is one of the constituents of over all environmental pollution. The menacing proportion in which it is growing in our environment these days is becoming a matter of concern for all of us. It has been established that excessive noise is not only adversely effecting the health of human beings but is also a health hazard to all living beings. Even the non living things are not left unaffected by high intensity of noise. Noise pollution is a pollutant in which man has added to his miseries himself, by not properly understanding the effects which noise has on public health. The necessity of a healthy environment is well recognised for human health and happiness. Galbraith has rightly said, "In the world into which economics was born the four most urgent requirements were food, clothing, shelter and an orderly environment in which the first three might be provided". It is true it is difficult to have an orderly environment in the technologically advanced modern age.

Noise as pollutant has become; a great nuisance

these days. It is spreading so fast that it has started polluting the environment of the society. It is a shadowed public enemy whose growing menace has increased in the modern age of industrialisation. Noise is unwanted or undesired sound. Pollution is 'any man induced change in the environment as a result of his activities, which has a measurable and generally detrimental effect upon the environment'. Pollution from a legal point of view is the wrongful contamination of the atmosphere, or of water or of soil, to the material injury of the right of an individual. Thus, noise as pollutant produces contaminated environment which affects adversely the health of a person and produces ill effects on living, as well as on non-living things.

**Sound and Noise**

Sound is the form of energy giving the sensation; of hearing and is produced by longitudinal mechanical waves in matter including solid, liquid and gas and transmitted by oscillation of atoms and molecules of matter.

Although a soft rhythmic sound in the form of music and dance stimulates brain activities, removes boredom and fatigue, but its excessiveness may prove detrimental to living things.

Noise is an unwanted sound without agreeable musical quality. Thus, otherwise sound and; noise can be taken to mean the same thing but in considering our acoustic environment, we must differentiate between these two terms. It is only when the effects of a sound are undesirable that it may be termed as noise. Encyclopaedia Britainica defines.

In acoustics noise is defined as any undesired sound. According to this definition, a sound of church bells may be music to others. Usually, noise is a mixture of many tones combined in a non-musical manner.

***Encyclopaedia Americanna defines it as:***

Noise by definition is unwanted sound: What is pleasant to some ears may be extremely unpleasant to other, depending on a number of psychological factors. The sweetest music, if it disturbs a person who is trying to concentrate or to sleep, is a noise to him, just as the sound of a pneumatic riveting hammer is noise to nearly everyone. In other words, any sound may be noise if circumstances cause it to be disturbing.

Pollution from a legal point of view, is generally speaking, the wrongful contamination of the atmosphere, or of water, or of soil, to the material injury of the right of an individual. Thus, noise as pollutant produces contaminated environment which becomes a nuisance and affects the health of a person, his activities and mental abilities. So noise pollution is unwanted sound 'dumped' into the atmosphere notwithstanding the adverse effects it may have on living and non-living things.

The scientific community has already accumulated considerable knowledge concerning noise, its effects and its abatement and control. Noise control also has a long history. The Ramans, perhaps, enacted the first prohibitory noise law when by a popular decree, chariots were banned from the streets at night.

**Sources of Noise Pollution**

The sources of noise pollution can be divided into two categories:

**Industrial Sources**

In industries noise is a by-product of energy

conversion. Cotton, mills, foundries and many other industries where big machines are working at a high speed have high noise pollution which requires our urgent attention in its minimisation.

**II. Non-Industrial Sources**

Non industrial noise pollution sources can further be divided into the following categories.

(a) *Loudspeakers:* One of the common factors creating noise pollution is indiscriminate use of loudspeakers. In India no function or ceremony is complete without a loudspeaker which has all the characteristics of creating of public nuisance. The recent spate in 'Jagrans' and 'Jagratas' and completion of these religious ceremonies in temple or private, which go on for whole nights, marks the unabated use of loudspeaker without a moments pause. The situation at the time of Ram Lilas or electioneering is much more shocking. One is forced to hear the discourses whether one likes it or not. Most people bear it and are reluctant to lodge a complaint for fear of unpleasant neighbourhood; relations and fear; of bringing the wrath of the users on them. Is this competitive religious fervour from every place of worship necessary? Is it not tortureous for a person who wants to rest and sleep, for a student who wants to study? The agony of a patient in such situation is beyond imagination.

(b) *Automobiles:* Automobiles constitute the largest single group of noise menace. In a city, 60 to 70 per cent of noise may come from road traffic. Slow speed of five to ten kmph during peak hours increase the emission rate of atmospheric and noise pollution. In India in cities road lengths are much less than desired, being 7 per cent of the total area, instead of 20 to 30 per cent, so the vehicle densities become alarmingly high resulting in increased noise level.

Western countries have been trying to alleviate the problem by manufacturing quieter motor vehicles, building roadside sound barriers, insulating adjoining buildings, going in for more expansive tunnels restricting operation of noisier traffic to certain hours or zones. Automobiles contribute toward noise pollution because no regulation is observed in blowing of horns and use of defective silencer pipes. It has become a fashion to remove silencer pipes from motor cycles and scooters by some miscreants. Undesirable noise produced by their vehicles causes people annoyance.

(c) *Trains:* In India steam engines are commonly used by railways which produces a lot of noise. The impact of noise pollution by trains is maximum in those areas where railway tracks are situated in residential areas, with the introduction of fast trains, the noise pollution has been substantially increased.

(d) *Aircrafts:* Higher the speed of an aircraft, the greater is the noise pollution. The supersonic aircrafts have added more noise specially so for living beings who live near aerodromes. The noise of these planes can break windowpanes, crack plaster and shake buildings. By these effects of noise one can easily understand the effects of such noise on human body. Major cities around the world have banned flights at night to prevent citizens having to put up with the deafening roar of jets. As our geographical position does not permit banning of night international flights, we will have to wait till we can enforce this legislation.

(e) *Construction Work:* Demolition and construction for urban renewal and expansions always make the urban man a victim of noise pollution. During demolition of old sites and construction of new; buildings huge machines which produce a lot of noise are being commissioned and it has become a common scene in every big city where construction work is in progress. Construction work may be unavoidable for social needs but the resultant noise produced may be hazardous for the health of the persons subjected to noise.

(f) *Projection of Satellites in Space:* A new source of noise pollution is Satellites, projected into the space with the aid of high explosive rockets. Application and use of these rockets produce deafening noise at the time of 'lift off' a Satellite. Tons of TNT and other explosives are used in these operations which create noise pollution as well as air pollution.

(g) *Radio, Microphones:* Radio and microphones can cause noise pollution if they are switched on with high volume. Present day interest in Western music and dance by youngsters with high volume causes noise pollution.

**Measurement of Noise Level**

Noise levels are measured in decibels. One decibel is the threshold of hearing. 30 decibels denotes the whispering range, 50-55 decibels may delay or interfere with sleep, 60 decibels is the level of normal talk; 90-95 decibel may cause irreversible changes in the automatic nervous system; 150-160 decibels prove fatal to some animals. The indecibls noise quantum in the noisiest area of some of the cities in our country are Delhi (89), Calcutta (87), Bombay (85), Madras (89), Cochin (80), Madurai (75), etc. One can easily dicern the high noise level in their cities. Situation though in small cities may not be so alarming at present but can become so if timely remedial measures are not planned. The trauma of noise pollution may prove fatal.

Noise researchers say that noise levels in excess of 90 decibels (unit for measuring noise intensity) for continuous periods can cause loss of hearing. A single exposure of 150 decibels is known to cause permanent injury to the ear's internal mechanism. In cities like Bombay, Calcutta, and Delhi, the average noise level has been found to be between 65 and 90 decibels. The effect of environmental noise on the foetal development during pregnancy has been the subject of research. It is found that constant exposure to noise between 110 and 120 decibels can produce narrowing of vision, vertigo and distruption of equilibrium in the unborn baby keeping the present developmental pace into account it is observed that the noise level too doubles its present value every six years. This is turn is affecting the hearing capability of the populace. By 2000 A.D. it is possible that no one above the age of 10 will have normal hearing capability.

**Effects of Noise Pollution**

The environmental deterioration may be caused by single or interactive effects of the several sources of environmental pollution like air, water, soil, noise, etc. These various sources of pollution are bound to have damaging effects on the behaviour of human beings and some of the effects are so serious that they create risk even for their survival. Frightened by these serious consequences psychologists, scientists, administrators and legal experts have started studying the effects of these pollutions on different and various aspects of human functions in a systematic and empirical manner. Noise is one of the main pollutants of the environment causing various hazardous consequences for human life. Noise not only impairs our sensibility to auditory stimuli by masking effects, it has other consequences too.

Researchers have proved that a loud noise during peak marketing hours creates tiredness, irritation and impairs brain activities so as to reduce thinking and working abilities. Noise pollution was previously confined to a few special areas like factory or mill, but today it engulfs every nook and corner of the Universe, reaching its peak in urban areas. Industries automobiles, rail engines, aeroplanes, radios, loudspeakers, taperecorders, lottery ticket sellers, hawkers, etc. are the main ear contaminators of the city area. The regular rattling engines and intermittent blowing of horns emanating from the caravan of automobiles do not allow us to have any respite from irritant noise even in suburban zones. However, the noise pollution's most apparent victims today are the residents of neighbourhood near larger airports. The introduction of jet planes have considerably increased their misery.

Its general effect on human beings is that it causes disturbances in sleep which lead to other side effects. It has auditory effects like loss of hearing. Broadly speaking it has:

**Physiological Effects**

This form of environmental degradation has implications for health as serious as air or water pollution. It can change man's physiological state by speeding up pulse and respiratory rates. It can impair hearing either permanently or temporarily; millions of industrial workers are threatened with hearing damage. Medical evidence suggests that noises can cause heart attacks in individuals with existing cardiac injury, and that continued exposure to loud noise could cause chronic effects as hypertension or ulcers, and of course deafness. Audiograms of pop musicians show typical hearing loss in both the ears. For the connoisseur of this music, it takes 20 to 50 hours of quiet to make the loss of hearing caused by a four-hour-rock concert.

The adverse effects of noise pollution on the human beings are manifested through physiological indications such as loss of hearing, occupational deafness and noise induced diseases. Some empirical research conducted on pregnant female mice reveals that aircraft taking off which bring in 120 to 160 decibels caused miscarriages in them. If the findings on mice are made applicable to human beings, high noise is equaly capable of creating similar disturbances in human beings also.

Several birds have been observed to have stopped laying eggs. Apart from this animals change their places. It has been observed that less and less number of migratory birds move to a particular place if the noise level at that place increases. Prolonged chronic noise can also produce stomach ulcers as it may reduce the flow of gastric juice and change its acidity. It may lead to abortions and other congenital defects in unborn children.

**Psychological**

Many behavioural changes are recorded as a result of exposure to high noise in human beings as well in animals. Certain symptoms can be observed outrightly. The undesired sound may cause annoyance. Intolerable agony may result when the source of the sound is not; known. Interruptions in speech communications may impair performance, lead to errors and lower output and efficiency. Noise can cause tension in muscles, nervous irritability and strain. No doubt the noise reaction varies to large extent in different individuals.

### Behavioural Effects

By lowering down the auditory sensitivity of person, noise result in poor attention and concentration. It has been observed that performance of school children is poor in 'Comprehension' tasks where schools are situated in busy areas of a city and are subjected to noise pollution. Noise causes irritation which results in learning disabilities. Sudden noise distracts a person and can create nervousness in him. Housewives working in kitchen with all kinds of electric gadgets have been to get headache due to noise and vibrations of these gadgets. Perhaps, this may be the reason that our saints always selected place for meditation away from residential localities and free from noise.

It has been found that in the Toda of the Nilgris that hearing capacity of a 50-year old man is as good as that of a 20-year-old. A study by an organisation in Ahmedabad finds an increased incidence of birth defects, still births and unusually low weight among children born to mothers living near airports, indicating that foetal development is adversely affected if the mother is subjected to continuous noise stress during pregnancy.

### Personological Effects

If the injurious effects of noise tend to persist for long, they cause stable maladaptive reactions in the individual disturbing his total personality make up. The lowered performance level among children may develop a feeling of inadequacy,lack of confidence, poor perception of one's own self which may jeopardize their optimal personological development as a growing child. Once the feeling of ineptness, worthlessness and inadequacy are developed by a child in the growing age, its disastrous effects are not going to be removed easily without leaving their marks behind.

According to Prof Mathur people suffering from insomnia, fatigue, hypertension, blood pressure and deafness were all showing symptoms of living in noise polluted areas. A bus or truck without silence may emit noise up to 110 decibels. Today noise is also being used as a third degree measure for interrogation. In the early 20th century a German Noble laureate, Robert Pock had predicted that a time is likely to come when man will die due to sheer noise.

**Other Effects**

Other effects of noise pollution may be summarized as under.

(a) There is a close relation between high noise levels and cardio-vescular diseases.

(b) Noise is one of the important factors responsible for hypertension.

(c) Sleep disruption due to noise can cause a wide range of problems like low quality work, irritability, aggressive social behaviour, etc.

(d) Noise is regarded as the genesis of stress diseases like ulcers.

(e) It also has connection with accidents.

**Effects of Noise on Animals**

Guinea pigs, rats, chinchillas, dogs and cats are favoured animals for noise effect studies. These animals are usually subjected to intense sound levels over relatively short periods of time instead of low levels over long time periods. From these tests, researchers have been able to observe damage done to animals hearing mechanism, blood chemistry, and stress response.

Most farm animals show good adjustments to noise, particularly the larger animals such as horses, cattle, and swine. However, poultry are not as adaptable to sounds, particularly to unexpected loud sounds. In fact, one of the earliest recorded official complaints related to aircraft noise occurred in 1923 when a farmer wrote to the Postmaster General complaining that low-flying aircraft were disrupting egg production.

There are practically no data available on controlled tests of wildlife under realistic, chronic exposures to noise. It would be useful in understanding the noise effects on wildlife, particularly with regard to animal panic behavior, affects on exploratory behavior, prey-predator relationships and mating signals.

**Effects on Non-living Thing**

The high intensity of noise effects non-living thing too, viz., buildings. The cracks in the city of the Recoco Church at Stemhausan were caused by some booms.

We can all narrate our experiences of the effects of noise. However, its interference with speech and sleep and the depreciation of the value of property for residential purpose near airports, and other noise prone areas is well known. Apart from this, people living in noisy areas may develop habit of shouting and even deafness. Aircraft noise can be the 'last straw' before onset of psychiatric disorders. There is a need to do medical research in India on effects of noise, so that its impact on human beings is well charactered and proved. Thus, the effects of noise are of concern for human beings, wildlife and non livings.

**Noise Control**

After the sound levels have been measured and the noise survey analysis has been completed, the data obtained can be compared with established criteria for potential hearing damage, for interference with speech, or for annoyance levels. Such a comparison will indicate the sources that should be controlled and the required degree of noise reduction. Noise exposure reduction may be achieved by the *application of engineering control techniques*, or by the *regulation of exposed personnel*. Engineering control methods are usually preferred. This refers to the alteration of design, changes in the mounting or operation of a noise source, or the construction of sound barriers, sound absorbers, etc.

Regulation of personnel implies reducing hazards by requiring the use of ear protective devices (such as earplugs), or by limitation of personnel exposure time. For purposes of illustration, this discussion will be limited to engineering control methods that are generally applied to industrial noise problems. Many of the basic control methods will, however have a much wider application.

The control of noise is most economically achieved in the planning stage. Avoiding a problem in this manner presupposes a knowledge of the noise characteristic of each machine or process activity while "on the drawing board". When noise potential information is available, or can be obtained with judicious investigation, it may applied to engineering design, process development, or plant layout. At this time, equipment or process noise specifications can be stipulated to suppliers. The design of quieter equipment by means of plant planning can be evaluated.

Equipment is designed primarily for the productive task to be performed. There are no "standard methods" by which a noise-free machine can be fabricated. Building in a quiet operation requires knowledge of the mechanics of sound generation and the path of sound transmission. Regardless of the equipment involved, the fundamental mechanics of sound wave generation may be classified as either a vibrating surface or as turbulence in a fluid medium. Generated sound may be transmitted by several paths other than direct radiation to the atmosphere. A moving part may set up; vibrations in some adjacent solid that could, in turn, become a noise generator. Likewise, air-borne sound waves may induce vibrations in solid surfaces that could, reradiate noise into the air. The control method applied will be dependent upon the paths of "noise flow", and the sound energy being carried by each path. Application of fundamentals regarding sound generation and transmission could prevent a noisy design. A similar application of fundamentals to plant layout could avoid objectionable noise. Even with the most careful design, certain equipment may create an objectionable condition. Isolation of the potential problems during the planning stages could prevent costly renovations.

Noise problems resulting from equipment that has been installed and in operation may be solved by the substitution of quieter equipment, or the substitution of quieter process activities and quieter materials. A few examples of these methods include: the use of a centrifugal fan, where practical to replace an axial flow fan (such as a propeller fan which produces, more high frequency noise); the use of welding to replace a noisier riveting process; the substitution of rubber tires to replace steel wheels on vehicles.

Engineering control methods may be imaginatively applied to reduce noise by making some modifications to an existing noise source. The fundamentals of sound generation and transmission are again applied. Control of a vibrating surface can be achieved by a damping or isolation of the forces responsible for the vibration. A common example is the isolation of vibrating parts by the use of resilient materials. Using springs or rubber cushions to isolate vibrating equipment from their mounting surfaces will reduce the noise that result from the continuous impact of vibration.

Relatively large vibrating surfaces, such as sheet metal sections, are frequent noise sources. Control may be achieved by reducing the vibrational response by such means as damping the member, improving its support, or increasing its mass. This type of member may be subject to resonance (vibration at its natural frequency). In this condition the driving forces are "in phase" with the members vibration and large surface displacements are common. Damping of surface vibrations is achieved by dissipating vibratory energy, thereby reducing the amplitude of motion. This may be achieved by covering the surface to the member with felt, mineral wool, etc.

It is frequently more feasible to reduce motion by improving the support of a vibrating member. This condition may occasionally be resolved by improving the fastening of the member to adjacent parts. Where this is not sufficient, the use of angle iron, cross members, or additional bracing may be required. These procedures will also serve to increase the mass and stiffness of a member, thereby reducing vibration; or shifting frequency to a less troublesome range.

A stream of flowing air, or some other moving fluid can create excessive sound levels due to turbulence established in the ambient atmosphere. The venting of exhaust gases, or the effluent from a high pressure nozzle are two common examples of this type of noise source. In many situations, the use of a muffler is the most convenient method of noise control. Several types of commercially available mufflers are commonly used. A dissipative muffler installed in a gas flow pipe or duct makes use of absorbing materials to attenuate sound. For hot or corrosive gases, stainless steel wool or synthetic fibres may be used as sound absorbing material. Ventilation ducts, which are frequently noise transmission paths, can also be lined with acoustical material.

High velocity jets are frequently found in air ejection systems for removing parts from presses. A high frequency noise usually accompanies this process. However, since the kinetic energy of the jet is used to provide an ejection force, the use of a muffler will not be acceptable. The velocity of the jet may be reduced if it is found that sufficient force can be achieved by moving the nozzle closer to the target and/or more accurately aiming the jet. Lowering the velocity could substantially reduce noise generation. For example, if the distance between the nozzle and the target were cut in half, the velocity could be reduced by 30% without reducing the ejection force on the target.

Reducing the velocity by this amount could lower the sound level by as much as 10 dB. In certain circumstances, complete conversion to a mechanical ejection system will be required to resolve the problem.

Exhaust streams and gaseous jets frequently sounds that are highly directional. Significant sound level reductions can be achieved by proper positioning and venting of these sources in a direction away from environmental occupants. Any obstruction in the path of a gas jet stream will create increased turbulence, and result in higher noise levels. Where some degree of obstruction is required, the streamlining of sharp edges and abrupt bends will help to reduce turbulence and lower sound levels.

Noise reduction at the source is not always possible. Once the sound has become airborne, attenuation may be achieved by some modification of the sound wave before it reaches the ear. This implies absorbing or dissipating sound energy along the airborne transmission path. The construction of a barrier or enclosure will confine most the sound energy to a limited volume. Efficiency of attenuation of the airborne sound is dependent upon the properties of the enclosure material (primarily the mass per unit area) and upon the sound frequency. The "acoustical opaqueness" of enclosure materials is described in terms of transmission loss (TL) expressed in decibels. This represents the ratio of acoustic energy transmitted through a barrier to the acoustic energy incident upon its opposite side. Transmission loss increases with increasing surface weight and with sound frequency. The use of transmission loss data reported in the literature will permit an estimation of the resulting noise reduction through walls, barriers or enclosures.

The greater the degree of enclosure and the more air tight its various members, the greater will be the noise reduction. For example enclosure walls constructed of materials impervious to air flow, the use of seals to insure tight-fitting joints and the isolation of the enclosure from vibration will improve noise reduction. A lining of some absorbing material on the inside of the enclosure may also be beneficial. Requirements of the process or noise source will frequently divide the extent of enclosure that is feasible. Where necessary, provision must be made for product flow, equipment operation, maintenance, or process venting.

Many undesirable noise situations result from the reflection of sound waves off walls, ceilings floors, or any smooth impervious surface in the environment. Surfaces may be covered with sound absorbing material in order to improve this situation. If all surfaces were perfect sound absorbers (having an absorption coefficient of 1.0) reflected sound; would be eliminated; and the reduction in intensity would follow the free filed, square of the distance relationship. This means a reduction of 6 dB with each doubling of the distance from the source. It is apparent that, although treatment with sound absorbing material can reduce noise in a general area, it cannot improve conditions close to a noisy source. Ventilation ducts and conveyors may be internally lined to attenuate sound transmission via this common route.

**Noise Pollution: Some Legal Perspectives**

The concern of the government for providing clean environment through environmental policy, planning and management has been very deep and sincere since 1970s. This is very clear from the national plan documents. The management of environmental despoilation was for the first time clearly provided in the Fourth Five Year Plan (1969-74). This plan highlighted the environmental issues in the following words

"It is an obligation; of each generation to maintain the productive capacity of land, air, water and wild life in a manner which leaves its successors some choice in the creation of a healthy environment. The physical environment is a dynamic complex and interconnected system in which action in one part affects others. There is also the; interdependence of living things and their relationships with land, air and water. Planning for harmonious development recognizes unity of nature and man. Such planning is possible only on the basis of a comprehensive appraisal of environmental issues particularly economic and ecological. There are instances in which timely, specialised advice on environmental aspects could have helped in project design and in averting subsequent adverse effects on the environment, leading to loss of invested resources. It is necessary, therefore, to introduce the environmental aspect into our planning and development. Along with effective conservation and rational use of natural resources, protection and improvement of human environment is vital for national well-being.

Since then the environmental matters have assumed significance and the Sixth Five-Year Plan (1980-1985) also attached importance to the subject of environmental conservation and control. Perceiving the problems of environment and its impact on national development, environmental management has come to occupy a place of priority at the hands of government. That the government is seized of the matter is clear from the fact that now a separate Department of

Environment (DOE) has been created to tackle the ecological crisis and problems. Some States too have also set up their own department of environment.

**Legal Control of Noise**

Pollutions are writ large on the fields, factories and facets of all forms of life. These are destroying the ecosystem, and consuming the qualities of environment. Hence, there is a growing need to regulate pollutions through law, so that the environment is saved from being polluted beyond repair, in the interest of mankind.

Many countries have enacted specific legislations to control noise pollution. For example in England there is Noise Abatement Act, 1960, of which Section 2 provides that loudspeakers shall not be operated (a) between the hours of nine in the evening and eight in the morning for any purpose; (b) at any other time for purpose of advertising and entertainment, trade or business. There are exceptions of course prescribed in the act.

The US, Noise Pollution and Abatement Act, 1970 is an important legislation for regulating control and abatement of noise. Under this law the environmental protection Agency, acting through the office of Noise Abatement and Control, holds public meeting in selected cities to compile information on the noise pollution. In some US States, environmental rights have been embodied in their constitution. As early as 1948, in Charles Kovacs V. Albert Cooper, AW city ordinance prohibiting the operation upon the streets of sound amplifiers or other instruments which emit loud and raucous noise and are attached to vehicles operated or standing upon such streets was challenged. A conviction for a violation of this ordinance, affirmed by the State appellate courts, was further affirmed by a majority of the Supreme Court of the United States as against the objections that the ordinance is lacking in definities and that it infringes upon the constitutional right of free speech. The majority agreed that sound amplification in streets and public places is subject to reasonable regulation, and least did not disagree that an ordinance prohibiting the emission of loud and raucous noises does not go beyond reasonable regulation. Even in Japan there is acute awareness about environmental management under the Anti-pollution Basic Law. Even a small country, like Israel, has taken initiative by enacting legislation to control anti-pollution activities.

In India it is important to note that there is no law which exclusively deals with problems of noise and its control. Though the Indian Constitution embodies provisions in Articles 39 (e), 47, 48-A and 51-A (6), Article (e) states that the health and strength of workers, men and women, and the tender age of children are not abused. Article 47 enjoins upon the State to raise the level of nutrition and the standard of living and to improve public health. Since 1976 Article 48-A provides:

The state shall endeavour to protect and improve the environment and to safeguard the forest and wild life of the country.

**Article 51A (g) reads:**

To protect and improve the natural environment including forests, lakes, rivers and wild life, and to have compassion for living creatures.

There are other micro provisions and enactments regulating loudspeakers in the States of M.P., Rajasthan and Bihar. Section 3 of the Bihar Act provides restrictions against the use and play of loudspeakers. It reads:

**No person shall use and play a loudspeaker**

(a) Within such distance as may be prescribed from a hospital building in which there is telephone exchange; or

(b) Within such distance as may be prescribed from any education institution maintained, managed, recognised or controlled by the State Government, or University established under any law for the time being in force, or a local authority, or admitted to such university, or any hostel maintained, managed or recognised; by such institution when such institution or hostel is in the use of students.

Section 6 provides that the cognizance of offence under the Act would be on a complaint made by, or at the instance of, the person aggrieved by such offence or upon a reporter in writing made by any police officer. Apart from this the Indian Penal Code also has provisions to regulate pollution on the ground of nuisance. The Motor Vehicles Act, 1939, also provides certain restrictions on trucks regarding noise by use of horns. Krishna Iyer J. rightly says:

"Local Boards Act if democratically implemented, Town Nuisance Acts and the Police Acts, if promptly and punitively used, the Criminal Procedure Code and Indian Penal Code if socially activated and proscriptively popularised, almost all nuisances, widely defined to include pollution on land, contamination of water, noise aggression and noxious discharges into the atmosphere, in short biosphere molestation, could be interdicted. But environmental protection is not the executive's cup of tea nor that of judiciary."

Law is too tame to halt the politics of pollution. The pity is that at the level of implementation, the law is suspended animation for extraneous reasons. Recently a landmark verdict of the Himachal Pradesh High Court shows a ray of hope for the people who have been victims of civic apathy and administrative neglect. It decides through a Public Interest Litigation suit on a significant aspect of the attitude of public functionaries and their duties. It said:

"When loudspeakers are allowed to disturb the neighbourhood, it is the duty of the police and the Deputy Commissioner to take appropriate action against those creating a nuisance and not wait for the suffering public to protect to the police.

The environmental (Protection) Act, 1986: One Step Forward, Two Steps Backward: The Environmental Protection Act, 1986 came into force from 19th November 1986 to commemorate the memory of Mrs. Indira Gandhi, who was globally recognised as a dedicated environmentalist. Her speech at the U.N. Conference on the human environment at Stockholm in June 1972, remains as one of the most important on the relationship between environment and development from the perspective of underdeveloped nations.

The Act purports to inculcate environmental ethics in every citizen and to take appropriate steps for the protection and improvement. The new Act for the first time attempts to lay down comprehensive law on environment and goes beyond the scope of the water and air pollution Acts passed in 1974 and 1981 respectively. Section 2 of the Act defines environment to include water, air and land and the interrelationship which exists among human beings, other living creatures, micro-organisms and property.

So this definition covers a much broader area than pollution.

But this Act has been linked to a cobra that is seemingly fierce. It raises its hood and hisses menacingly but if you prise its jaws open, you will discover it has no venom in its fangs. This is rightly so because it restricts the right of environmentalists to go to court on ecological issues. Under the Act a person cannot directly file a petition in court on questions of environment. One is required to give a notice of not les than 60 days to the Central Government of his intention to make a complaint. One can only go to the court if the Government does not act on the notice during this period. It may be submitted that such provision impede rather than serve the cause of environmental protection. Section 6 of the Act provides rules to regulate environmental pollution, wherein Section 6 (2) (3) provides for the maximum allowable limits of concentration of various environmental pollutions (including noise) for different areas.

Another important feature of the new Act that became law on May 23, is the inclusion of a section on hazardous industries and environmental disasters. A 'hazardous' substance according to the new Act, is 'any substance or preparation which, by reason on its chemical or physio-chemical properties or handling, is liable to cause harm to human beings, other living creatuses, plants micro-organisms, property or environment.

Stringent measures have been provided to check hazardous pollution. Section 8 of the Act states clearly that 'No person shall handle or cause to be handled any hazardous substance except in accordance with such procedure and after complying with such safeguards as may be prescribed'. And section 6(f) empowers the Central Government to make rules for the procedure and safeguards for the prevention of accidents which may cause environmental pollution and for providing remedial measures for such accidents. Moreover, it is now mandatory for a person responsible for the discharge of any hazardous substance in excess of the prescribed norms to immediately inform the concerned authorities and to render all possible assistance. Earlier there was no such responsibility enjoined upon him.

The new Act has some drawbacks. For one, all power and authority is vested in the hands of the Central Government. There is no free delegation of powers to the State Governments. Even the authority or authorities constituted to implement the Act are subject to the supervision and control of the Central Government. Excessive centralisation could become a major hurdle for the efficient execution of the provisions of the Act.

Another drawback is in Section 24(2) of the Act where any act or omission constitutes on offence punishable under this Act and also under any other Act when the offender found guilty of such offence shall be liable to be punished under the other Act and not under this Act. This is an anomaly, as most of the offences committed under the new Act would also constitute offence under the Water and Air Acts. And the penalties provided for in these Acts are less stringent than those permitted by the new one. Hence offenders against the common provisions of Acts, under 24(2) of the New Act, would have to be punished whereas according to the earlier Acts, could get away with a lighter punishment. This lacuna in the Act needs to be removed.

Yet another apparent drawback pertains to the

issue of industrial pollution. Section 6(e) of the Act empowers the Central Government to make rules for 'the prohibition, restrictions on the location of industries and the carrying on of processes and operations in different areas.This is no doubt a welcome provision as upto now there were only guidelines issued in this regard.

What is lacking, however is a specific mandatory provision for the industry to prepare and submit to the concerned authorities a suitable environmental impact assessment report before the location chosen is approved.

It may be stressed here that the role of environmentalists and other voluntary bodies instead of being curtailed should have been enhanced under the Act. Public interest litigation cases will suffer a set back under the provisions of the Act. In fact the act has taken one step forward and two backward. The Act smacks of over-centralization of powers in Central Government which is not conducive to the protection of environment.

This makes abundantly clear that the pathology of legal importance in overcoming pollution is not that we don't have enough laws but most laws with police powers bark but do not bite. Public interest litigation with broadened rules of locus stand for initiating such actions is the right answer to such problems. Environmental litigation as included in public interest litigation is now new, even to India.

What is warranted for is uniform law for controlling noise pollution so that the society is freed from this hazard. Legislation must envision positive action and not stop with punishing the offenders. A national environmental policy and ecological plan must be promoted by law so that man and nature love and live in harmony. There is no escape from pollutant. It reaches us through the air we breathe, the water we drink, the food we eat and the sounds we hear. In view of the alarming proportions of noise, and its impact, what is called for is 'noise control' through technology, determination of administrators, public and judiciary. Because only a healthy environment provides a healthy body and a healthy mind. Man is maker of his own environment, and he should not add to his miseries by himself polluting it. People should be made increasingly aware of the part; that natural environment plays in determining the quality of their lives. Participation in environmental protection should be properly campaigned and projected through local organisations, conservation councils and commissions. Environmental law societies should be encouraged like SOCLEAN (Society for Clean Environment, Bombay) in all parts of the country. Public Interest litigation is the right measure and a positive help by judiciary for curbing noise pollution and for promoting the cause of environmental quality throughout the country. Apart from this, law schools, should include environmental law in the curriculum of the law course so that the law students are acquainted with the problems relating to environment. Krishna Iyer J. rightly says:

'The Constitution commands us, the law forbids us. Regulatory legislation to control environmental pollution is a must since the rule of law must defend the rules of life. And life will survive only if the biosphere is safe. But law is paper tiger unless education makes society militantly aware of the risks of pollution. Let us not pollute or degrade God given environment when he ensures ecological balance.

The Chief Justice of Himachal Pradesh High

Court as he than was along with Thakur J. on 28th September 1983 delivered a judgment of utmost importance which has by far gone unnoticed. The judgment shows the rare legal acumen which possessed the concern of the court for the enforcement of citizens rights and to direct the authorities to perform their statutory duties.

Here the Court admitting a public interest litigation suit held that it is the right of every affected person to come to the court to ask for relief and wherever the court is convinced of the cause the court will say it is the duty of the court, to issue appropriate orders.

Besides other issues the important; point which came up before the court for its consideration was the use of loudspeakers. The petitioner alleged that the loudspeakers are being used indiscriminately and so loudly that these have become a nuisance. The court held "that though the conditions imposed on the use of loudspeakers are laudable indeed. They are mostly being observed in the breach. We may also record that in our court room may times we have to stand the noise of the loudspeakers used by religious institutions. It appears that some religious institutions are bent on instilling the fear of God in the society by using the loudspeakers in such a way that these can be heard over by the maximum people of the town. It seems that in their zeal the religious institutions are following the principle of home delivery services.

The court further held: Despite clear provisions of law unfortunately the loudspeakers are being misused. Obviously the authorities concerned fail to enforce the law. The offences being cognizable, the police is required to take cognizance of the same the moment the offence is being committed within their presence.

Since the offences of misusing the loudspeakers can only be detected by the sense of hearing, it is the duty of the police to take action the moment a case is made out. The police is not required to wait for a citizen to come forward to lodge a report of any cognizable offence before they swing into action. We find that the District Magistrate has pointed out that the citizens are required to lodge a report with the police and if the police does not take any action, they should go to the Magistrate and file a complaint since the Magistrate will take cognizance of the same under section 190 of the Code of Criminal Procedure. It only shows that the District Magistrate does not attach the importance to this problem which it deserves. Had he applied his mind, we would not have failed to realise that having recourse to the Magistrate may be an exercise in futility since the loudspeakers are mostly used only for a limited time by different persons. For example, it is not practicable for a student who is preparing; for his examination or for a person who is sick to move the court for the remedy. For such like situation it is expected of the police to swing into action quickly and not to wait for individuals to lodge the reports. We are also conscious of the fact that most of the individuals tolerate this noise pollution because of others. The Deputy Commissioner has not specifically denied the allegations of the petitioners that patients in the Ripon Hospital are disturbed by the loudspeakers and the bands which are played at odd hours near the religious places on the Cart Road. His only reply is that according to the conditions the permission is automatically null and void if there is any hospital within 100 meters. We expect the Deputy Commissioner to show more responsibility. He should apply his mind before granting permission for the use of loudspeakers. In other words, he should himself apply the conditions which are stated in the form by refusing to grant the permission, if the loudspeaker is to be used within 100 meters of a hospital, or of any school, college or University or Government offices during the working hours of the latter.

In India there is exclusively no uniform law for controlling noise pollution. Some countries have enacted specific legislations to control noise pollution. However, in India there are provisions in different Statutes which deal with noise pollution, some States in India do have law on the use of loudspeakers and its control but the laxity and lack of will on the part of administration, a total neglect by society, leaves the law only as a paper tiger.

As a result of rapid urbanisation, 156 million of India's population stay in cities. This figure is expected to escalate to 350 to 400 million by the year 2000 A.D. Delhi's population is expected to cross the 10 million mark by the vehicular population will double the present 13 lakh. According to an OECD (Organisation of Economic Co-operation and Development) study, in the absence of control measures, 53 per cent of Delhi's population will be affected by noise by 2000 A.D.

There is an urgent need to do research on the effects of noise on human health by medical profession and other research agencies. The medical profession should start a vigorous programme in environmental education so that the common man is made aware of the hazards of noise pollution Law and Society should join hands in making environment healthy. Because only a healthy environment provides a healthy body and a healthy mind, which is prerequisite for human happiness and progress.

# 4

# Nuclear Pollution

**Introduction**

Nuclear pollution is caused due to the addition, through activities of man, of ionizing radiations to the environment giving people an exposure to more of such radiations than they normally would experience.

Radiation is able to permeate the universe, the solar system, and the earth. The surface of the planet would get bombarded with radiation from the sun so intense that life would be in danger if it were not for the atmosphere that surrounds the earth. The atmosphere is able to screen out much of the sun's radiation, including most of that which would be lethal to life. Cosmic rays, coming from outer space at high velocity are able to strike the earth continually and penetrate deeply into the surface. Naturally occurring radioactive elements are found in the rocks, water, and air and in all living organisms.

Radiations enter the body on airborne, radioactive dusts and gases, or in food. The death rate from lung cancer in the infamous uranium mines of Joachimsthal, Bohemia was 30 times normal in the 1930s. Radon, a radioactive decay product of uranium and radium, has been a chemically inert but radioactive gas that did not dissipate readily in the mines, which have been worked for lead, cobalt, and arsenic, and the frequency of "lung diseases" among the miners was noted more than 400 years ago. As radiation damage in a cell has been most serious when it affects the generic chemicals that are critical in cell division, the symptoms of such damage appear earliest among tissues whose cells divide most frequently. Tissues forming the walls of the stomach and intestines and the tissue that makes white blood corpuscles in bones have such cells. It has been not surprising, therefore, that early symptoms of radiation damage involve the stomach (nausea and vomiting) and intestines (diarrhea), as well as a drop in the supply of white blood cells. Strontium 90 (half-life 27.7 years) has been dangerous because it has been a bone-seeker,being chemically similar to calcium, a component of bone. Cesium 137 (half life, 26, 6 years) is chemically similar to sodium, an element present as the sodium ion in ordinary table salt. Ions of cesium 137-are easily distributed throughout the entire body, including the gonads.

From a biological point of view, radiation effects are also classified as somatic or genetic. Somatic effects have beem the effects on the body itself and have been of direct concern to the person exposed to the radiation. Genetic effects have been those imvolving mutations of the chromosomes or genes in sex cells; they pose a potential hazard to the descendants of the exposed person and are of concern to the whole society. Genetic effects have been of grave concern only in persons who have been not yet past their reproductive years.

We are lacking a basic understanding of the biological action of ionizing radiations despite a great deal progress in research. Most experience have beem carried out using laboratory animals, and comparison with humans has been difficult to make. Some data could be accumulated over the years from human exposures to large radiation doses obtained accidentally or from nuclear explosions. Table 11 lists the approximate short-term effects that might be experienced for whole body adiation exposures over a short period of time. A whole-body exposure of 1 rad would mean an average absorption of 100 ergs/s over the whole body. Table 10 refers to immediate effects, and there is no guarantee that recovery might not be followed many years later by other effects-such as greater incidence of cancer than that occurring in unexposed persons. As different persons differ in their sensitivity to radiation, as in their sensitivity to infections and numerous other things, the doses corresponding to different effects will vary widely from person to person.

Some of the incoming radiation gets trapped by the earth's magnetic field. The far-reaching portion of the atmosphere, called the magnetosphere, is having an area of high-energy radiation called the Van Allen region. It is roughly doughnut shaped, extending from about 500 miles above the earth at the equator to an altitude of about 40,000 miles. The Van Allen region was at first thought to present a major obstacle to space travel, but further studies revealed that astronauts would pass through the zones of high radiation quickly and that precautions could be taken to disallow excessive exposure.

We have now learned how to produce radiation in many forms, and our exposure to the normal level of environmental radiation that remained relatively constant for millions of years has been greatly increased. We are being bombarded with man-made radiation from x-ray machines, radioactive fallout from nuclear explosions, effluent from nuclear power plants, radioactive materials and other radiation sources in research laboratories, industrial plants and hospitals, wastes from reprocessing nuclear fuels, and from mining and processing radioactive products.

Extremely small amounts of radiation can be profoundly injurious, although different forms of radiation are having different biological effects. Visible light and infra-red heat rays are kinds of radiation that are generally beneficial, but some of the more potent forms of radiation, for example, x-rays, destroy parts of living cells and tissues. As these highly energetic forms of radiation tend to split substances, including matter, into ions, they are known as ionizing radiation. The various forms of ionizing radiation differ greatly in their penetrating power and in the intensity of their injurious effects.

Sun has been the source of spectrum of radiations like radiowaves, infra-red, ultraviolet, x-rays, gama and cosmic rays. In addition, radioactive isotopes give off subatomic particles like protons, neutrons, electrons and helium nuclei called x-particles in the process of decomposition from an unstable state to a stable state to a more stable condition.

**Kinds of Radiation**

Radiation is a form of energy which may be transferred from one body to other through empty space.

Radiation has been of two main types. Sunlight—the form of radiation people are most familiar with—is an example of the type called electromagnetic radiation. It covers a broad spectrum of radiant energy and a wide range of physical and biological effects. The other type, called particulate radiation, consists of parts of atoms which are travailing at high speeds, often with tremendous energy. Both electromagnetic and particulate radiation are having great biological importance and account for some of the most serious environmental pollution problems.

**Electromagnetic Radiation**

Electromagnetic radiation consists of a broad spectrum of energy having the nature of light. Radiowaves, infra-red rays, visible light rays, ultraviolet rays, x-rays, and gamma rays have been all part of the electromagnetic spectrum. All the different kinds of electromagnetic radiation have been nothing more than light rays of different wavelength and frequency. The fact that our eyes have been sensitive only to visible light, actually a very small part of the light spectrum, has been enough to remind us how different the world would look if, through some accident of nature, we could also see x-rays or radio-waves.

The energy of electromagnetic radiation gets transmitted in packets called photons. All photons, of whatever part of the spectrum, travel at a speed of 186,282 miles per second, the speed of light. During their flight though space, they seem to vibrate, so that the rays have a wavelike quality.

The energy of a photon is considered as being inversely proportional to its wavelength. It means that the shorter the wavelength, the more energy there is in the photon. Therefore, as the wavelength of the electromagnetic spectrum decreases, the radiation has been said to become more energetic. For example, the visible rays of sunlight have been more energetic than radio rays of longer wavelength, and ultraviolet light is more energetic than visible light. X-ray having a wavelength of 0.1 nanometer (1 $A^0$) possess 3,000 times as much energy as ultraviolet rays having a wavelength of 300 nanometers. As the wavelength decreases and the energy of the radiation increases, the potentialities to disrupt living tissue would become very great. The electromagnetic wave has two component parts - an electric and a magnetic moving in phase, but in direction $90^0$ from each other much like two vibrating strings—one going up and down and the other back and forth, superimposed on each other. The wave carries no net electrical charge and no net magnetic moment but because of the components which can interfere or react with electric or magnetic fields, it can lose or gain energy (i.e., change frequency).

**1. Microwaves**

These are radiowaves in or near the extremely high frequency, or shorter wavelength, range. Hence, they are electromagnetic radiations in the more energetic portion of the radio wave spectrum. Microwave energy has been much too low to disrupt living tissues by ionization. Instead, the energy gets absorbed as oscillation energy and is converted to heat. This makes it possible to use microwaves for cooking. But the same radiations have been capable of heat injury to tissues in the body when exposed to the radiations, thus making it necessary to carefully maintain the door seals of microwave ovens. While microwaves can bring about bodily damage with heat, most of our discussion will be dealt with the more energetic radiations that can disrupt tissues, including cellular structures, by ionization, or splitting, of molecules.

**2. Ultraviolet Rays**

Ultraviolet light is electromagnetic radiation just beyond the shortest wavelength of violet light to which the eye has been sensitive. It is having profound physical and biological effects. Though shorter in wavelength than the most energetic rays of visible light, ultraviolet light rays are longer than the average for X-rays overlapping the shorter and more energetic x-rays.

Ultraviolet light covers from 2-180 nm in vacuum and 180-400 nm for far and near UV of electronagnetic spectra. Ultraviolet light is produced in abundance by the sun and can be generated artificially by the carbon arc or the mercury vapor arc. Ultraviolet radiation penetrates human skin only superficially, probably to a maximum depth of 1 millimeter, which, however small is enough to cause serious injury. Ordinary glass about 2 millimeters thick filters out (absorbs) ultraviolet rays shorter than 310 nanometers. It has been fortunate that the air, smoke, and dust filter out much of the ultraviolet radiation from the sun, especially the shorter wavelengths, for otherwise the damaging effects would drastically change life on earth. The shorter wavelength ultraviolet radiation near the x-ray region can bring about damaging ionization.

**3. X-rays**

These are produced when a fast moving electron is suddenly accelerated/decelerated, releasing electro-magnetic (em) radiations called bremsstrahlung. Another mode of production has already been discussed before.

X-rays include a broad spectrum of wavelength. Therefore, the energy, penetrating power, and disruptive force of x-rays differ greatly depending on the wavelength. An "average" x-ray is having a wavelength of approximately 0.1 nanometer, a distance about equal to the diameter of a molecule of water.

The discovery of x-rays was done by German physicist Wilhelm K. Roentgen in 1895. He was working on the luminescence of chemicals produced by cathode rays when he noticed a strange phenomenon occurring some distance from the tube. A sheet of paper coated with a luminescent substance (barium platinocyanide) was glowing. As the cathode tube was blocked off by cardboard, it was not at first apparent that the rays from the cathode tube could cause the luminescence. Roentgen reasoned that there would be some kind of penetrating radiation that was invisible to the eye. He experimented and reported that the radiation could penetrate thin layers of metal. Because he had no idea of what the strange radiations could be, he called them by the traditional symbol for the unknown, "x" rays. The name continued to be used after the nature of x-rays became known. For his excellent work, Roentgen received the Nobel Prize in 1901, the first year of its award.

**4. Gamma Rays**

Gamma rays are similar to x-rays, with which their wavelengths overlap, but because most gamma rays are of shorter wavelength, they generally have higher energy and are more penetrating than x-rays. Short wavelength x rays, sometimes called hard x-rays, may be identical to gamma rays at the longer wavelength end of their spectrum. Gamma rays are having wavelengths of about 0.1 nanometer and shorter. Regardless of the wavelength, the term gamma rays is used only in reference to electromagnetic radiation that is emitted from the atomic nucleus of a radioactive element. For example, some atoms of radium, uranium, and plutonium give off gamma rays.

These are produced by change in the energy levels of nucleus of atoms. Wavelength of this radiation varies from 0.0003 to 0.1 nm. The common sources are cobalt-60 and cesium,-137. The radioisotope of cobalt has been the source of two monochromatic gamma beams with energies of the gamma quanta 1.17 and 1.33 MeV. The half-life is 5.3 years. Cobalt-60 has been the most widely used source of gamma rays for radiochemical and radiobiological studies.

Cesium-137 is a radioisotope having a half-life of 30 years and gives both electrons and gamma radiations. Gamma ray energy is about 0.662 MeV. It is also widely used as a source of radiation for therapy, radiobiological studies etc. Other sources of gamma rays have been (a) the spent fuel rods from nuclear reactors, which are actually blocks of uranium, and (b) products of fission.

**Particulate Radiation**

Particulate radiation gets produced when one or more of the various components of the atom, for example, an electron, a proton, or a neutron, have been emitted from a radioactive element. Such particles may get ejected from the atom with great force, sometimes along with gamma rays as the atom gives up its energy, or in technical terms, as it gets decayed to a less energetic form of matter. The time it takes one-half a the nuclei of the atoms in a sample of radioactive element to decay is termed as the half-life. But whether the radiation from nuclear disintegration is electromagnetic or particulate, the emanations have been so energetic and forceful that they can do great damage to living cells and tissues.

**1. β-Particles**

Beta radiation is made up of high-speed electrons called beta particles. They get emitted by many of the radioactive substances during their decay to more stable states. An electron is exceedingly small; it weighs only 1/1,835 as much as a proton. When an electron is rejected at high speed from the atom of a radioactive element, it ionizes substances it collides with in its path. Thus beta particles damage tissue.

But they are less penetrating than x-rays or gamma rays. Carbon 24, phosphorus 32, hydrogen-3 (tritium), and strontium-90 are examples of beta emitters.

Electrons with energy range 3 to 100 MeV can be obtained from the various types of accelerators, such as the Van-de-Graaff generator. Such installations can produce electrons with energies of upto $10^2$ MeV and dose rates of the order of $10^{10}$ roentgen (R) per hour. Aside from these generators, there are many other radiation sources such as betatrons, linear accelerators, etc.

**2. Alpha Particles**

Some radioactive isotopes emit fast moving particles, called alpha particles, containing 2 protons and 2 neutrons. This is equivalent in make-up to a helium nucleus. Since alpha particles contain no electrons, they are positively charged. Alpha particles are less penetrating than either beta particles or x-rays and gamma rays. They could be stopped by a sheet of paper. Alpha radiation, therefore, is dangerous only when the emitter is in direct contact with tissue, such as when inhaled or ingested. But in such cases the alpha particles are readily absorbed by tissue, so the danger is very great. Radium-226, uranium-238, and plutonium-239 are examples of radionuclides that emit alpha particles.

**3. Protons**

Protons are 1,835 times heavier than electrons. While the energetic electron (beta particle) drives into tissue like a tiny particle of sand, the proton lumbers along like a rock, knocking off pieces of atoms and molecules at it goes. The proton does not penetrate as far as an electron of the same energy, but it causes more disruption in a small area. It has been said to leave a heavier track. The effect has been similar to that of an alpha particle.

**4. Neutrons**

It is a kind of nucleon, i.e. one of the building-blocks of atomic nuclei. The neutron has a mass of 1.8 x $10^{-30}$ kg or, in the conventional units, 939.57 MeV/$Cc^3$, in and is about 1 fm in diameter. It is similar to the proton but is electrically neutral, hence the name, and can be understood as a compound of three particular kinds of quark.

Neutrons are uncommon in nature because they decay radioactively with a half-life of 11 minutes, yielding protons symbolically:

$$n \rightarrow p + e^{-} + \text{antineutrino.}$$

This is a form of beta-decay. The practical importance of neutrons comes from the fact that a neutron can stimulate a nucleus of a fissile substance such as plutonium to undergo fission, with a release of energy. Neutrons can also be absorbed by 'fertile' nuclei (e.g. thorium) to produce additional fissile material. Neutrons will also be absorbed by the metal lithium with the production of the fusion fuel tritium. In fission reactors, neutrons are both a product and a cause of fission, a fact underlying the chain reaction concept. However, to ensure proper reactor operation, the flux of neutrons has to be carefully controlled-by moderators which slow neutrons down, by absorbers which remove them from circulation and by reflectors with the effect of preventing neutron loss by turning them back into the reactor core. Neutrons are copiously given off by the cores of fission reactors and can be a radioactive hazard through neutron activation; on the other hand their application in fuel-breeding makes neutrons very precious in the sense that they can be used to multiply man's energy resources many times over through the use of thorium and the heavy form of uranium ($^{238}U$).

**5. Energetic Neutrons**

Neutrons are having approximately the same mass as protons but have no charge. They are obtained from either a neutron beam or from the atomic disintegration of one of the radioactive isotopes, for example, uranium-235. Fast neutrons have been four or five times more lethal to mice than x-radiations. Neutrons shot from uranium-235 was the basis for the chain reaction of uranium in Fermi's famous "pile" under the bleachers of the stadium at the University of Chicago in 1942, where experiments on the first atomic bomb were conducted.

Major sources of neutrons have been nuclear reactors for irradiation of chemical and biological specimens. These give a mixed emission, composed of a stream of neutrons (thermal and fast) and gamma rays. The intensity of the stream of thermal neutrons has been of the order of$10^5$ to $10^9$ R/hour.

Neutrons are also available from neutron generators and cyclotrons in which low Z nuclei ($H^2$, $H^3$, $Hc^4$,$Be^9$) are bombarded by high velocity and positivcly charged particles.

In order to get beams of heavy particles of very high energy, the various types of accelerators are employed. For example, the Van-de-Graaff accelerators are able to produce protons of an energy level ranging from 1 to 10 MeV. Small cyclotrons provide beams of accelerated protons, deuterons and alphaparticles with energies of upto 10 MeV and with a penetrating ability (in biological tissue) of the order of a fraction of a millimeter. Beams of greater penetrating power (with energies of several tens of MeV) are having path length of the order of several centimeters. In order to get beams of even higher intensity, linear accelerators can be used in series with a cyclotron. In that case, the object of irradiation must be properly cooled so as to disallow the over heating, and consequently undesirable side reactions.

**6. Cosmic Rays**

Cosmic rays are energetic sub-atomic particles, from the sun and outer space. They strike the earth at high velocities, some of them penetrating several thousand feet of solid rock. Cosmic rays are part of the background radiation that affects all living organisms. The dosage from cosmic rays at sea level is estimated to be equal to about 0.001 roentgen per day. The amount of radiation that can be tolerated by humans without detectable injury is about 0.01 roentgen per day. Because the difference between background cosmic radiation and detectable human tolerance is small (about ten-fold), cosmic rays probably have significant effects on cells and tissue, no doubt causing mutations and other chromosomal aberrations. It has been estimated that cosmic rays account for about one-fourth of the mutations caused by radiation.

**Types of Radiaton on the Basis of Ionisation**

In its broadest sense, radiation is energy being propagated from one place to another through space. There are maiuly two types of radiations:

1. Non-ionising radiations
2. Ionising radiations

Ionizing radiations are the potent sources of energy. Their toxicity is almost 100 million times more than that of cyanides and other reagents.

**1. Non-Ionising Radiations**

Radiations of shorter wavelength but having greater energy may be able to harm the microorganisms but are able to injure only the surface tissues of higher plants and animals. These radiations are also known to increase the rate of mutations. Although some plants are able to absorb maximum radiation in the ultraviolet region, they can grow well in green houses where practically all the ultraviolet light has been filtered out. This shows that ultraviolet light is not required for plant growth.

In biological systems, there are certain enzymes which make the cell to repair or eliminate the genetic damage. The damages caused due to ultraviolet rays are repaired by activation of an enzyme upon exposure of dimers- having DNA to strong illumination. These damages could also be repaired in dark in which there occurs the removal of a part of DNA carrying the dimers through the activity of enzymes.

**2. Ionising Radiaction**

The radiation concerning pollution is ionizing radiation, which is the radiation of sufficiently great energy to ionize atoms and molecules. An atom gets ionized when it gains sufficient energy for one or more of its electrons to get separated from the atom; ionization of a molecule might yield two charged fragments, such as $H_2O \rightarrow H^+ + OH^{-*}$. If the fragments are uncharged, then they are referred to as "free radicals," as in the example $H_2O \rightarrow H + OH$. Free radicals are generally very reactive chemically.

The most important types of ionizing radiation from the standpoint of pollution (as opposed to the standpoint of the scientific researcher interested in new knowledge of the fundamental properties of matter) have been alpha, beta and gamma radiation:

1. Alpha radiation is made up of energetic alpha particles. An alpha particle is a $He^4$ nucleus, which has two protons and two neutrons which are bound

together into a very stable particle by the "strong nuclear interaction". Alpha particles carry a positive electric charge whose absolute magnitude has been twice that of the electron's charge and have been capable of interacting strongly with ordinary matter (including living tissues) by electromagnetic interactions.

2. Beta radiation is made up of energetic electrons (or their antiparticles, the positions). These particles are produced in the beta decay of nuclei, which also produces—antineutrinos (or neutrinos) which interact only very weekly with matter and have been of no importance as radiation As beta particles being electrically charged, they can interact strongly with matter by the electromagnetic interaction.

3. Gamma radiation are made up of very energetic photons, i.e., very short wavelength electromagnetic radiation. Despite being uncharged, photons are able to bring about very strong electromagnetic interaction with matter.

Also, many other types of ionizing radiation are encountered less frequently, but they can also be dangerous to living systems—protons, neutrons, deuterons, and other nuclei of sufficiently great energy, for example. the particular importance of alpha, beta and gamma radiation lies in the fact that they are produced in the radio-active decay of certain atomic nuclei, i.e., in the spontaneous or induced disintegration of these nuclei.

**Plutonium**

Plutonium, chemical symbol Pu, is an artificially produced, radioactive, chemically and radiologically highly toxic metallic element in the actinide series, which also occurs in minute quantities as a result of natural disintegration processes in nature. It is formed in nuclear reactors from uranium (U) by the following process: U-238 +neutron → U-239; U-239 disintegrates by the emission of beta radiation into neptunium-239 (Np-239), and this decomposes in the same way into Pu-239. Plutonium has the atomic number 94; and from it the isotopes Pu-239 to Pu-246 are known, of which Pu-244, with $8.2 \times 10^7$ years, has the longest half-life. The other isotopes also have a very long-life. The technically most important of them, Pu-239, has a half-life of 24,400 years.

As it is highly fissionable by slow neutrons, Pu-239 is used as nuclear fuel and to make atomic bombs (the atomic bomb that was dropped on the Japanese city of Nagasaki on August 9, 1945, was a plutonium bomb). It is easier to extract than uranium (U-235), and its critical mass amounts to no more than 8-16kg, compared with a critical mass for uranium of 50 kg. Pu-239 is now obtained in large quantities from nuclear reactors, especially from fast breeder reactors.

Radiologically, plutonium, which burns in the air to form aerosols capable of penetrating into the lungs, is one of the strongest poisons known. When they get into the human organism, quantities of as little as one ten-millionth part of a gram are sufficient to cause cancer, for plutonium emits alpha rays of limited range which are capable of very high radiation doses locally. Investigations by Tamplin and Cochran in 1974 serve to demonstrate that the margins for plutonium hitherto considered tolerable, and the legal limits prescribed by law for plutonium contamination, are in fact too high by a factor of 300,000.

Plutonium, is given off into the environment in limited quantities by reprocessing of nuclear fuels. There is likely to be in the future a substantial increase in the transport of plutonium to reprocessing centres by road and rail, so that the possibility of releases of plutonium by leakage or owing to terrorist activities cannot be ruled out. If a transport vehicle containing 25 kg were to overturn in an accident, the plutonium so released could cause up to 250 billion cases of lung cancer.

From the same quantity of plutonium, given sufficient technical knowledge, an atomic bomb could be manufactured. The dangerous nature of plutonium, having regard in particular for the long life-span of its isotopes, is one the main arguments against using nuclear fission as a source of energy.

**2. Radioactivity in the body**

The human body contains very small quantities of the radioactive isotopes carbon-14 and potassium-40. The $^{14}C$ originatesin the atmosphere and results in a dose of one mean per year in the soft tissue. Potassium 40 (40K) is naturally-occurring (half-life 1.27 x $10^9$ years) and contributes about 20m rem per year to the gonads.

A significant contribution to the radioactivity in the body comes from the gaseous decay products of the uranium and thorium radioactive series, namely radium and boron. these gases diffuse from the rocks and soil and are present in easily measurable concentrations in the atmosphere. They are breathed by man along with their decay products and are also taken up by plants and animals with the result that most foodstuffs contain measurable amounts of natural radioactivity. Of ordinary foods, cereals have a high radioactive content while milk produce, fruit and vegetables have a low content. The intake of this type of radioactivity varies greatly with diet and is thought to contribute an average gonadal dose of about 4 mrem per year in the British Isles.

### 3. Cosmic Radiation

Another important source of natural radiation has been cosmic radiation. Cosmic rays have been high-energy charged particles (mostly protons) of extra terrestrial origin. They have been capable of producing other energetic radiation by collisions with oxygen and nitrogen nuclei in the atmosphere. Table 3 lists some products of cosmic rays. The radiation dose obtained from cosmic rays has been found to be greater at greater heights in the atmosphere and may amount to about 10 rem/year above the atmosphere. Natural radiation from cosmic rays will be higher at higher elevations and higher in a jet aircraft at 10 km than on the ground.

Products of cosmic radiation, their half-lives, and their concentration in disintegrations per minute per cubic meter of air in the lower troposphere. The concentrations refer to those resulting from cosmic radiation and do not include the effects of nuclear weapons tests.

| *Isotope* | *Half-life* | *Concentration* |
|---|---|---|
| $H^3$ | 12.3 years | 10 |
| $C^{14}$ | 5760 years | 4 |
| $Be^7$ | 53 days | 1 |
| $S^{35}$ | 87 days | 0.015 |
| $P^{32}$ | 25 days | 0.015 |
| $P^{31}$ | 14.3 days | 0.02 |

Cosmic radiation reaches the earth from inter stellar space and from the sun. It is composed of a very wide range of penetrating radiations which undergo many types of reactions with the elements they encounter in the atmosphere. The atmosphere acts as a shield and reduces very considerably the amount of cosmic radiation reaching the Earth's surface. This filtering action means that the dose rate at sea-level is less than at high altitudes. For example, the mean dose rate to the gonads at sea-level, at the equator, has been measured as 23 mrem/year, while the equivalent dose rate at 10,000 feet, was 56 mrem/year. The average dose rate in the British Isles from cosmic radiation is about 50 mrem/year.

One very important radionuclide arises from the interaction of neutrons in cosmic radiation with nitrogen in the upper atmosphere to carbon-14 as follows:

$$^{14}N(n, p)^{14}C$$

The carbon-14, which has a half-life of 5568 years' diffuses to the lower atmosphere where it may become incorporated in living matter. Similarly, small concentrations of other radionuclide such as $^{3}H$ (half life 12.26 years) $^{36}Cl$ (half life 3.08 x $10^5$ years) and $^{41}Ca$ (half-life 1.1 x $10^5$ years) are maintained in the lower atmosphere by cosmic ray reactions. They are much less important than $^{14}C$.

**4. Background Radiation**

Natural atomic radiation emitted by the trace concentrations of radioactive materials in the rocks and soils of our planet as well as the cosmic rays coming from the sun and outer space.

Levels of background radiation are usually given in millirems (mrem) per year, the unit of dose equivalent, but sometimes the nearly equivalent unit of the millirad (mrad) per year is used. Although the background radiation varies widely from place to place in the world. For most settled areas of the United States it averages 140 mrem/year (Table 1). In the Travancore region of eastern India, active monazite sands produce a background of 200-2600 mrem/year. Certain states in Brazil have the same sand, the background radiation there ranges from 500 to 1500 mrem/year.

About 50 of the approximately 350 isotopes of naturally occurring elements are radioactive. On an average, the top 6 inches of soil on our planet contains a gram of radium per square mile. One of the "daughters" of radium, created as radium decays, is the radioactive gas, radon. Granitic rocks contain radioactive materials. (In fact, granite may someday be processed to obtain nuclear fuels). We take in radioactive materials in the air we breathe, the food we eat, and the fluids we drink.

**Table 4**

*Background Radiation*

*Total Exposure of the Average Person in the United States to Ionizing Radiations.*

| *Background radiation* | *Amount (mrem / year)* | *Voluntary exposure* | *Amount (mrem / year)* |
|---|---|---|---|
| Cosmic rays | 50 | Jet flight, cross | |
| Ground (6 hours/day) | 15 | country | 4 |
| Building materials | | Watch dials | 2 |
| (18 hours/day) | 45 | Television | |
| | | (1 hour/day) | 5 |
| Air | 5 | Medical and | |
| | | diagnostic | |
| Food and water | 25 | X-ray | 55 |
| Total | 140 | Total (est) | 65 |

An individual's exposure to radiations in excess of background comes from watching television, travelling in a high altitude jet (giving more exposure to cosmic rays), wearing a watch with a luminous dial, and being X-rayed at the office of the dentist or the doctor. These more recent additions to our total exposure to radiations are part of the reason for the concern of many radiation biologists. Moreover, nuclear power plants add radioactive materials to the water, the air, and the food chain. A nuclear power plant run without incident may add 5 mrem/year to the atmosphere and 0.05 mrem/year to the water used for cooling within the plant. In relation to the background radiation, this has been judged to be insignificant unless one ignores the concentrating effect of food chain on certain radioactive isotopes or the cumulative effect of several nuclear plants clustered in densely populated areas.

**5. Man-made Sources of Radiation**

The early experiences of man-made sources of radiation involved X-rays and various uses of radium. As early as 1896 a letter appeared in Nature describing the effects of repeated exposure of the hands to X-rays and during the next 15 years many more cases were reported. These cases arose both from experiments with X-ray sets and also from their use in various treatments. By 1911, Hesse had studied the histories of 94 cases of tomorrows induced in man by X-rays, of which 50 cases were among radiologists.

These studies illustrated the early types of damage produced by X-rays but gave no indication of the longer-term effects. For some types of damage, such as skin cancer, there is a latent period of between $10^{-30}$ years and some radiologists observed malignant skin changes as late as 25years after discontinuing fluoroscopic examinations. By 1922 it was estimated that more than 100 radiologists had died from occupationally-produced cancer. Similarly, it has been estimated that the death-rate from leukaemia among early radiologists was about nine times that among other physicians.

Other studies have shown that the average life-expectancy of the pioneer radiologists was reduced by approximately 203 years compared with physicians in general practice. Using results from animal experiments on the relationship between lifeshortening and amount of irradiation, it is estimated that the total dose received by the average radiologist from 1935-1958 ranged between 400-800R. The results of this study have been used by the International Commission on Radiological Protection (I.C.R.P.) to set the occupational exposure limit of 15 rems per year for most organs of the body.

Experience was also gained early in this century of the effect of internal dose from various nuclides such as radium ($^{226}$RA, half-life 1622 years), mesothorium ($^{228}$Ra, half-life 5.8 years), radiothorium ($^{229}$Th, half-life 1.91 years), and their daughter products. Even earlier, it had been recognized that there was a remarkably high incidence of lung cancer among the minors in the Schneeberg cobalt mines of Saxony and the Joachimsthal pitchblende mines in Bohemia. Eventually, this high rate of lung cancer was shown to be due to radiation from the daughter products of uranium, namely $^{226}$Ra, $^{222}$Rn, $^{218}$Po etc. These mines contain large concentrations of uranium.

During the second and third decades of the twentieth century there were many cases of overexposure to radium. A considerable number of these arose from the use of radium as a therapeutic agent. It was administered for a large variety of diseases ranging from arthritis to insanity!

The most serious overexposures to radium occurred in the radium-dial painting industry in the United States. Most of the persons employed were women and they had the habit of 'pointing' their paint brushes with their lips. Many of these women probably ingested tens or even hundreds of microcuries of radium. It is not known how many radiumdial painters actually died from the effects of radiation damage.

The maximum permissible body burden of $^{226}Ra$ set by the International Commission on Radiological Protection (I.C.R.P.) is 0.1 Ci which corresponds to an average dose to the skeleton of about 30 rem/year. The study of people exposed to X-rays and radium is continuing to try to establish the degree of risk from acute or chronic exposures to radiation.

**6. Radiations from Medical and Dental Exposure**

The public are getting a slowly increasing dose of ionizing radiations from medical and central X-rays. As Table 5 indicates these man-made exposures are considered the largest source, above the background, of ionizing radiations of man. Most of the exposure has been for diagnosis ("checkups"). Unfortunately, most of the X-ray machines used for medical purposes have been operated by physicians and chiropractors who lack special training in health physics, radiation protection, or the use of their machines. Because radiologists, the specialists in medical X-rays work, tend to be employed at hospitals and clinics with high workloads, the majority of X-ray actually obtained are taken under the supervision of radiologists. The nonspecialist, has been more likely to have old equipment, to understand its use less well, to allow greater exposures than necessary (sometimes because of a cheap camera and a slow lens), and to be less careful in focusing the beam. These generalizations are also applicable to the field of dentistry as well. There should be laws relating to periodic inspections of equipment and to requtirements for the training of physicians and dentists have been exceedingly lax. One is greatly concerned about pelvic X-rays given to women of child-bearing age. The embryo and fetus have been particularly sensitive to damaging radiations. the International Commission of Radiation Protection and Measurement (ICRP) urged in 1962 that except in emergencies, any radiological examinations of women in this age group be limited to periods when pregnancy is most improbable, which the ICRP defined as the 10-day period following the onset of menstruation.

It has been estimated that 75-90 per cent of the total exposure of the population from medical uses of radiation comes from the diagnostic use of X-rays. The most critical regions of the body are the bone marrow, the gonads and the foetus. The bone marrow is the site of the primitive bloodforming cells and so irradiation of this region could lead to the induction of leukaemia. Irradiation of the gonads is important because of the possibility genetic damage. Irradiation of pregnant women should be very severely limited because it can lead to physical and mental deformities of the child.

In most countries the average dose to the population from therapeutic radiology is much less than that from diagnostic radiology. Although quite large exposures may be used in certain treatments, only a small number of people are involved.

**7. Uses of Radioisotopes**

Radioisotopes are used in medicine to give a means of tracing the path and location of specific chemicals in the body. Since radioactive isotopes are chemically identical to stable isotopes of the same element, they will follow the same path and be concentrated to the same degree as the non active isotopes in the body. By counting methods the location of the active, and hence of the ordinary non-active, isotopes of the element may be determined.

**8. Radiation from Television sets**

For a person viewing television at a distance of 6-7 feet, the average absorbed dose to the gonads, according to data quoted by Morgan, is probably much less than 1 millirad/year, provided there is a safety glass in front of the screen and the set operates at less than 25 -kilovotts. For a 25-kilovolt set without the glass, the absorbed doses at 6-7 feet could be as high as 1.1 millirads/year to ovaries and 7.5 millrads/year to testes. (Since for radiation protection purpose the units of the rad and the rem are taken as roughly equivalent, this dosage, as seen in Table 1.3 exceeds exposures caused to the general population by currently operating nuclear power plants and is of the same order of magnitude as exposures being currently received as the result of fallout from nuclear weapons testing). This dosage or any widely distributed dosage is a serious matter because of the large mumber of people involved, therefore, the potential changes in the gene pool, as well as the statistical increase in incidence of leukeamia and other diseases induced by radiations.

**9. Radiation from Nuclear Power Plants**

The contribution of nuclear power plants to man-made radiation is appreciably small. Radiation standards set by the U.S. Atomic Energy Commission (AEC) formerly allowed a maximum dose of 500 mrem/year at any point on or beyond the boundary of a nuclear power plant, which would several times the usual natural background; in 1971 this limit, got reduced to 5 mrem/year for light-water-cooled nuclear power reactors. Operating nuclear power plants (except for less than half a dozen older plants) in practice emitted no more than a few per cent of the old 500 mrem/year limit , and in the great majority of cases less than 1% of it. Hence the actual exposure of a person remaining around the clock at the least favorable location (in the direction of the prevailing winds at the plant boundary) would be no more than 5mrem/year, in accordance with the new limits. It is thus believed unlikely that anyone in the U.S. is receiving even 1 mrem/year radiation exposure from nuclear power plants—an amount which is approximately equal to that received from increased cosmic radiation exposure during a single transcontinental jet flight.

**10. Radioactive Fallout from Nuclear Weapons**

Radioactive fallout from nuclear weapons tests has been of grave concern to scientists. When we test a nuclear weapon in the atmosphere, there occurs local fallout of radioactive fission products over the immediate area for about a day, then worldwide tropospheric fallout for about a month from fission products released into the troposphere, and stratospheric fallout worldwise for many years thereafter. This fallout has been easily detected at scientific laboratories which are located around the world and its study has greatly increased our knowledge of worldwide atmospheric transport processes. The rapid transport of fallout from one hemisphere to the other has been indicated by the detection within 22 days at 34° north latitude of radioactive I131 and Ba 140 from a French nuclear test at 21° south latitude.

So as to be of concern to man, a radioactive

fission product must be produced in sufficient quantities, possess a sufficiently long half-life (which is as short as 8 days in the case of $I^{131}$), be transferable to man, and remain in the body long enough to do damage. The most hazardous radionuclides turn out to be those listed, $C^{14}$ has been produced naturally in the atmosphere by cosmic radiation and has been a constituent of all living tissues; weapons tests through 1975 increased the amount in the atmosphere by 80 to 100% but circulation in the biosphere will be able to reduce this to about 3% by 2040A.D.unless atmospheric testing continues. $Sr^{89}$ has been similar to (but much shorter-lived than) $Sr^{90}$ as a pollutant that gets discriminated against biologically but has been still of great importance. $Sr^{90}$ and $I^{131}$ both reach man through cow's milk, the $Sr^{90}$ going into bones and $I^{131}$ into the thyroid gland. $Cs^{137}$ reaches human tissues through ingestion of milk and meat but has a biological half-life (i.e., half-life inside the body) of only 70 to 140 days because of its fairly rapid removal from the body through metabolic action.

**Table 6**

*Radionuclides Important in Fallout*

| *Element* | *Isotope* | *Half-Life* |
|---|---|---|
| Carbon | $C^{14}$ | 5760 |
| Strontium | $Sr^{89}$ | 51 days |
| Strontium | $Sr^{90}$ | 28.9 years |
| Iodine | $I^{131}$ | 8.1 days |
| Cesium | $Cs^{137}$ | 30.2 years |

Due to the variations in half-lives (both physical and biological), and environmental happenings that flush materials into unpopulated domains (e.g., the ocean), these averages have been not very meaningful except that they represent doses, in comparison with those received from medical and dental exposures, that are small. (The current average per year is estimated as about 1.5 millirads). Further atmospheric testing, of course, would have pushed these figures higher and higher. And people near to or downwind from the tests - to say nothing of the victims of Hiroshima and Nagasaki—had been seriously harmed.

### 11. Peaceful or Constructive Uses of Nuclear Explosions

Another expected man-made radiation source in the future would be the "peaceful" or "constructive" uses of nuclear explosives for releasing natural gas from underground regions, for building harbors, or even for constructing a new sea level canal between the Oceans. Natural gas released with a nuclear explosion has been likely to get contaminated by radioactive gases such as $Kr^{85}$, for example, and large-scale construction processes may lead to radio-active contamination of air and water. Even the use of radio-isotopes for medical and scientific purposes involves some risk to the general population; the burnup of an orbiting SNAP (Systems for Nuclear Auxiliary Power) generator in the atmosphere in April 1964 led to detectable amounts of $Pu^{238}$ in rainwater. The risks involved in the peaceful uses of nuclear energy are outweighed by the benefits.

Danger of radioactive pollution principally lies in nuclear war and in testing of nuclear weapons. Both short and long term effects have far reaching implications. Since 1945, with the initiation of nuclear weapon testing and the disastrous results at Hiroshima and Nagasaki there has been considerable awareness of nuclear hazards. The US nuclear power submarine, lost in the Atlantic in 1963, releases a large quantity of radioactive materials in the environment. As such, even if nuclear weapons are not used during major wars, radioactive pollution of the ocean is a continuous threat to mankind. the total number of nuclear detonations up to the end of 1978 was 1,619 by the United States of America, 247 by the USSR, 29 by France, 24 by Peoples Republic of China and one by India (c.f Office of Public Affairs, USA). In the United States, it is true that even at the time of maximum testing the average fallout was 30 mrem, well below the guidelines of 170 mrem per year. But in spite of this minimal dosage the shortening of life span due to any exposure to radiation cannot be over looked.

The nuclear tests may be underground, underwater and atmospheric. The underground tests, as carried out by India, can be fully contained and generally result in cratering. The Plowshare Programme or Atomic Energy Commission,USA, recorded that cratering explosions may release 10% of the radioactivity into the atmosphere the ecological balance is usually not upset the danger is posed by the remaining 90% which may be leached and affect the underground water. It is quite likely that underground explosions may release radioactive contaminants as often noted in the burning of radioactive natural gas in the process of stimulation of its production from the geological strata of low permeability.

The classification of the underwater test is based on the criterion of depth under the surface of water and that of the atmospheric test on different heights from the surface of the earth up to the stratosphere. The principal threat to our ecosystem and human life lies in underwater and atmosphere tests.

Several peaceful uses of nuclear explosion have

been programmed, including the acceleration of gas and oil production and formation of under ground storage chamber for gas and oil. Others involve use in mining projects, refineries, excavations as well as in the construction of harbours and canals. Precautions are normally taken in such cases to contain hazards up to the level of zero discharge. Though paradoxical, the techniques of radiation and problems of industrial pollution.

**12. Nuclear Wastes**

The increasing use of radioisotopes and, more particularly, the development of the nuclear power industry results in an evergrowing quantity of radioactive waste. Continued dispersal of low and intermediate levels of radioactive waste to the environment means that members of the general population will receive an increasing exposure from this source. At present the contribution to the total exposure of the population from waste disposal is negligible but it seems certain to increase as more power stations are built. For this reason, very strict control is exercised over the dispersal of radioactive waste to the environment.

Nuclear wastes from atomic power plants comes in the form of spent fuel roads of uranium and the by-products such as plutonium. It is estimated that these can remain toxic to humans for over 200,000 years. Radioactive iodine, another waste product from power plants can cause cancer of the thyroid gland. Waste coming from the production of nuclear weapons produces radioactive strontium and cesium both of which are carcenogenic. These materials generate heat and penetratin radiations from centuries. The nuclear wastes such as contaminated dust, debris, clothing, industrial clothing trash, etc. When dumped into the disposal sites can defile the atmosphere and even seep into the soil and pollute the drinking water. It is, therefore, necessary, that wastes coming from power stations and defence establishments should be carefully handled, isolated, buried and protected.

**13. Fallout from Weapons Testing**

The nuclides of concern in radioactive fallout from nuclear weapons testing are similar to those arising from the operation of nuclear power stations. The most important radionuclides are strontium-90 ($^{90}$Sr, half-life 28.8 years) and cesium-137 ($^{137}$Cs, halflife 30.0 year). Strontium-90 concentrates in the skeleton and cesium-137 is distributed uniformly throughout the body.

Some of the radionuclides created during a nuclear test are injected into the troposhere (40,000-60,000 feet) and are carried around the earth several times. They gradually return to the earth over a period of a few years and consequently give appreciable doses to the world's population. The dose from nuclides injected into the troposphere reaches a peak shortly after each weapon test.

The fall-out from atomic explosion also cause pollution in the atmosphere. The radioactive dust that falls to earth after atomic explosion is called radioactive fall-out. These materials mix and interact with natural particulate materials in the atmosphere and increase the amount of man-made air pollution. The kind of radioactive fall-out depends upon the type of bomb.

Fall-out from weapons differs from atomic waste materials in that the radio-nuclides are fused with non-silica, and dust and whatever happenes to be in the vicinity to form relatively insoluble particles. The smaller particles adhere tightly to the leaves of the plants where they may not only produce radiation damage to plant's tissue but may be injected by grazing animals and then to their alimentary canal.

The fall-out can enter the food chain directly at the hrbivore or primary consumere level. In gereral, the total amount of radioactivity decreases with distance from a nuclear test. There are certain exceptions to this especially strontium-90 which reached a peak in wild animal populations 50 to 100 miles from 'ground zero' of the explosion.

The amount of all-out received in an area is roughly proportional to the rainfall. the quality of fall-out radionuclides that enters food chains and eventually becomes transferred to man depends not only on the amount received from the air but also on the structure of the eco-system and the nature of its biogeochemical cycles . In geeral a large proportion of all-out will enter food chains in nutrient poor environment.

Fallout radionuclides have been and are being passed on to man through the food chain, although concentration in human tissues are not generally as high as those in sheep and deer. Man is somewhat protected by the position in the food chain and by food processing and cooking which removes some of the contaminants.

**14. Radiations from Miscellaneous Sources**

The cigarette smoker, as usual, reaps special harvests. Besides cancer-inducing tars, his lungs are getting some cancer-inducing radioactive lead (at least 40 millirems/year for the pack-a-day smoker); it is roughly a third of what he cannot escape because of background radiation (which does not all go right to the lungs). There has been an estimated gram of radium, on the average, in the top 6 inches of each square mile of soil. When radium decays it produces radon, another radioactive substance when is a chemically inert gas, which mingles with the atmosphere. The decay of radon produces a radioactive isotope of lead that comes out of the air and deposits on foliage like tobacco leaves.

There have been also a number of other everyday sources of man-made radiation. Older-type luminous watch faces using radium could give local dose of up to 2 mrem/h, but tritium ($H^3$) is used today and the doses are only a negligible portion of the man-made radiation. X-rays from black-and-white television sets are contributing another millirem per year and large doses have been occasionally reported from other sources –10 mrem/h from houses built with radioactive stone,100 mrem/h from houses built with radioactive stone, 100 mrem/h from bathtubs glazed with uranium pigments, etc.

**Chernobyl Nuclear Mishap**

On April 28th, 1986, the melting of a nuclear reactor and its explosion had taken place at the Chernobyl nuclear power station, 80 miles north of Kiev in USSR. It is regarded the worst disaster in 32 years of commercial atomic power which has caused unspecified deaths, suffering and environmental and health damage. The dangerous radiation also affected neighbouring countries.

Around 2,000 people near the plant had been killed by causes which were ranging from the initial blast to lethal radiation and about ten thousand were evacuated from the endangered region. Radioactive gases and particles have spread over a vast section of the Soviet Union and water supplies for more than 6 million inhabitants of the Kiev area get threatened with contamination.

The most serious part of the nuclear accident was the radiation emitted from the reactor, which was then carried by winds to the neighbouring areas. In the first few hours of the disaster deadly form of iodine and cesium got released into the atmosphere along with other highly dangerous radioactive emissions. By the end of the week a large invisible cloud of radiation covered Eastern Europe and the shores of the Mediterranean sea. The people who have been actually in trouble have been those right at the site and local residents who risked exposure to extreme doses of radiation that could give rise to cerebral haemorrhage, nausea, vomitting and death within hours.

**The Most Dangerous Radioactive Pollutants**

Strontium-90, cesium-137, and iodine-131 have been by far the most serious pollutants in the general environment from fallout and as atomic wastes. A radioactive isotope is regarded to belong the "serious pollutant" category if it satisfies the following criteria:

1. It is formed in a high yield in whatever nuclear reaction makes it (fission or fusion).
2. Its half-life has been in the range of 1-1000 years. (If much shorter, it decays rather quickly; if much longer, its activity is low).
3. It is easily taken up by a living organism—either man or something in the food chain of man.
4. It is not eliminated from important body organs (i.e. it has a long biological half-life).

Iodine-131 (half-life 8 days) has been a serious pollutant in spite of its low half-life because it has been produced in high yield by fission (both in weapons and reactors), it is efficiently taken by cows from grass and hay and put into milk, and one tiny gland, the thyroid of man and animals, very effectively concentrates it.

Radioactive phosphate cesium and iodine-131 can also readily accumulate in plants and animals through natural food chains. However, in food chains involving arthropods radioactive isotopes of potassium, sodium, and phosphorus accumulate, but isotopes of strontium and cobalt do not.

A less common but locally important radioactive isotope classified as a serious pollutant has been plutonium-239.

*Strontium 90* normally, is formed in radioactive fallout, has a half-life of 8.1 years and behaves like calcium in biogeochemical cycles. Thus it gets absorbed by plants, ingested by animals and deposited in bone tissues close to blood forming tissue. Strontium-90 is able to concentrate in natural biological systems in following method: water → bottom sediments → aquatic plants → fresh-water clams → minnows and small fish → musk rats. It is experimentally proved that due to this food chain musk rats concentrate strontium-90, 35,000 times above the levels of the water in which they live. Grazing animals are able to concentrate strontium-90 by ingesting it through grass and forage, and it can then be passed on to humans through milk.

Its beta rays emerge with 0.55 megaelectron volt. In the environment it is concentrated in food chain of animals, being a bone-seeker. Strontium-90 ions are chemically similar to calcium ions, and at least on surfaces of bone tissue can replace some calcium ions.

Although isotopes are able to accumulate in human tissues as well as those of plants and animals, it could not be established so far whether current levels of isotopes in human tissues represent serious health hazards to man (*Southwick*, 1976: *Smith*, (1977). Some medical scientists such as *Gofman* and *Tamplin* (1970) and *Sternglass* (1972), however, feel that man's radiation exposure from artificial sources has been already sufficient to produce serious disease problems (leukemia and bone tumors), genetic damage, and infant mortality.

Tritium is an isotope of hydrogen having a mass of 3 units. It is an emitter of relatively low energy beta rays and has a half-life of 12.36 years. Its beta rays cannot penetrate intact skin, but tritium occurs as tritiated water (HTO) in which form it easily enters living organisms by food, drink, humid air, and even direct absorption through the skin. Tritiated water undergoes the same kinds of chemical reactions as ordinary water, providing a route for the distribution of tritium throughout the entire body. Some of it inevitably becomes incorporated into the basic genetic material, DNA, where it may cause molecular change leadings to tumors, cancer, or fetal abnormalities,

Most but not all of the studies involving exposure of experimental animals to ingested tritium, either as tritiated water or as tritiated DNA (the chemical of genes), have shown that tritium has no unique and special powers of damage compared with X-rays or gamma rays. A dose from tritium has the same radiobiological and radiation protection meaning as the same dose, with the same rate pattern, of X-rays or gamma rays. Therefore, tritium can cause cancer, and this effect has been demonstrated in laboratory mice exposed to the isotope. By the year 2000, assuming that nuclear power enjoys a heavy growth, the exposure of the general population to tritium released from nuclear power plants will average not more than 0.001 millirem/year. This may be compared with the current contribution of tritium to our backgroung radiation of 0.06 millirem/year. Consider also the roughly 120-140 millirem/year from watching color television, or the several millirem from exposure in a long jet flight or a high-altitude sking vacation. Average exposures, however, tend to hide what might be serious local effects near power plants or atomic fuel reprocessing plants. The maximum permissible concentration of tritium in water is 0.0003 microcurie per cubic centimeter. If the seven nuclear power plants planned for the shores of Lake Michigan are built, they will discharge tritium into the lake until by the year 2000 its concentration will be about 0.2% of this maximum.

Tritium is produced in nuclear reactors in a number of ways. Some comes from relatively infrequently occurring fission events—one tritium atom in every 10,000 fissions. Coolant additives (e.g, boric acid or lithium hydroxide) capture escaping neutrons and undergo subsequent nuclear reactions that release tritium. Light water (i.e, ordinary water) contains traces of deuterium, another isotope of hydrogen, which can be changed to tritium by neutron capture. (That is why tritium releases from heavy water reactors used in Canada and several other countries are much higher than from light water reactors in the United States) Compounds of lithium and boron are present in control roads, where they interact with neutrons to produce tritium.

Plutonium is a man-made, silvery, metallic element (atomic number 94). It forms as a bay-product from uranium-238 in burner reactors of nuclear power plants, and it would be continuously made in proposed breeder reactors.

Plutonium 239 is the most important of several plutonium isotopes. Thus far its chief use has been to make warheads of atomic bombs or the triggering devices of hydrogen bombs. The importance of plutonium-239 in civilian nuclear power is growing. In the Nuclear Energy report of Project Independence (November 1974). the Federal Energy Administration assumed that nuclear power plants would obtain plutonium fuel from the reprocessing of spent uranium fuels starting soon after 1977. Since a number of dangers exits, this projection might be optimistic.

When plutonium-239 enters the environment from weapons tests or accidental releases from reprocessing plants, it appears as plutonium oxide, a very fine dust that is insoluble in water Plutonium-239 emits powerful alpha and gamma rays and has a half-life of 24,360 years. Its alpha rays are about 10 times more harmful in lung tissue than X-rays or gamma rays. Plutonium-239 is probably the most poisonous substance known. Inhaled into lungs, it causes lung cancer. If it enters the bloodstream, it causes bone cancer.

The greatest danger of plutonium-239 lies in its potential as a fuel in atomic bombs. Every nuclear power plant produces this material as its uranium fuel 'burns'. Plutonium-239 can be separated from spent uranium fuel-about 25 kilograms of Plutonium-239 from one year of operation of a 1000-megawatt power plant. A simple implosion atomic bomb requires 4-8 kilograms of plutonium, and atomic bomb. When nuclear reactors are sold to countries wanting to join the "nuclear club", the countries may be expected to develop their own atomic bombs in 6-10 years. As more and more countries develop their own bombs, the possibilities that one will attempt nuclear blackmail against others increase.

The theft of plutonium between the time it is

produced in a reprocessing plant and the time it is loaded into a nuclear reactor is another cause of grave concern in government and environmentalist circles. With only modest scientific expertise, a terrorist group could make a small atomic bomb from stolen plutonium.

## Radiation Protection Standards

*Introduction*

Soon after the discovery of X-rays and radio activity a sizable number of workers started using ionizing radiations. Their number increased manifold after the start of nuclear energy programmes in different countries. Hundreds of thousands of works have been associated with this programme directly while millions of people in a number of countries have been affected in some way or the other by nuclear radiations. Right from the early days of discovery of ionizing radiations the deleterious effects of over exposures to radiations had been observed. Therefore, a need to prescribe some guidelines for the use of radiations was felt. It was only in 1928 that a unit of measuring the quantity of radiation exposure, the roentgen (R) was established.

There were obvious difficulties in formulating standards of radiation dose due to lack of information on the correlation between dose and the injury it produced in an organisms. For instance, it is very difficult to determine the limit of radiation below which no significant injury will result. No direct expermients can be conducted on human beings. Extrapolation of findings on animals is bound to be in error due to differences among animals of different species. Inspite of these handicaps a large amount of information on the radiation effects has become available from the use of X-rays and radium sources in radiology and cancer therapy, from victims of atomic blasts accidents at nuclear energy installations, and general population exposed to atomspheric tests conducted by the major and less major powers. Therefore, the radiologists and health physicists have formulated certain broad proposals which have been accepted as guidelines for radiation protection.

Some of the principles underlying these guidelines pertain to the recovery from the effects of small radiation doses. The body cells are replaced quickly or slowly depending on the nature of the tissue. Consequently a minor somatic injury is cured with time. However, the genetic effects are cumulative and there is no recovery from them. Secondly all living beings are exposed to cosmic rays and other ionizing radiations continuously. Radiation dose received from cosmic rays is 400 to 500μ Sv $y^{-1}$, another 900μ Sv $y^{-1}$ is received from terrestrial sources like residences, soil, clothes, etc., while another 200 to 300μ Sv $y^{-1}$ arises from sources inside the body namely, water, found, air, $^{40}K$, $^{226}Ra$, i$^{4}$C, etc., so that on an average a dose of 1500μ to 1700μ Sv is received annually from natural sources and the human body has adjusted itself to this without any apparent damage.

On the international level the only widely recognized authority for setting maximum permissible exposures to ionizing radiations is the International Commission on Radiological Protection (ICRP), founded in 1928. It was chiefly concerned with the safest possible uses of X-rays and radium consistent with the human benefits these were known to offer. From the agency's inception, ICRP policy has been to permit no necessary exposure. The problem has been in defining what is necessary. When long-term risks are impossible to assess by short-term experiments. How can we know when benefits outweigh risks? The benefits seen most obvious for medical uses, and therefore, the ICRP has allowed exposures in this area to be 30 - 100 times the levels deemed acceptable for the average person in the population at large. Giving the patient such exposures, of course, means extra risks to the medical personnel who handle this work or, in the case of radium, to people who mine and process radium ores. Thus the category of occupational exposure limits came into being.

Since its inception, the ICRP has been the one internationally recognized body which is responsible for recommending values of maximum permissible exposure to ionizing radiation. Most of the national legislation in many countries relating to exposure to radiation has been based on the recommendations of the ICRP.

The early recommendations of ICRP were concerned with protection against X-rays and radium. Some of the earliest recommendations were dealing with the length of time that a worker should get engaged on radiation work; these were:

(a) Not more than 7 h/day.

(b) Not more than 5 days/week.

(c) Not less than 1 month's holiday per year.

(d) Off days to be spent as much as possible out of doors.

The maximum permissible doses had been defined very loosely.

In 1950, the ICRP tried to extend its scope to deal with the many new problems which had resulted from the introduction of nuclear reactors. The maximum permissible exposure levels recommended by ICRP have changed considerably over the past four decades. All of these ranges have resulted in a decrease in the occupational exposure values. This continuous decrease in the maximum permissible exposure level has not been attributed to any positive evidence of damage to persons working within the earlier permissible dose levels, but rather to an increasing awareness of the uncertainties in much of the available experimental data.

**Table 7**

*Values of Maximum Permissible Occupational Exposure*

| *Dose rate* | *Date recommended* | *Comments* |
|---|---|---|
| 0.1 of an erythema | 1925 | Proposed by A. dose per year Mutscheller and R.M.Sievert. This corresponds to an exposure of ~30 R/year from 100 kV X-rays or ~70 R/year from 200 kV X-rays. |
| 0.2 R/day or 1 R per working week | 1934 | Recommended by I.C.R.P. |
| 15 rem/year, or approximately 0.3 rem/week | 1950 | Recommended by I.C.R.P. |
| 5 rem/year, or approximately 0.1 rem/week | 1956 | Recommended by I.C.R.P. |

Over the years many different expressions are being used to describe acceptable levels of exposure to radiation. The most important of these have been tolerance dose and maximum permissible dose. The term tolerance dose was having the unfortunate connotation that it seemed to imply a threshold dose below which no radiation damage would take place. On the basis of presently available evidence there has been no proof of the existence of a threshold dose for certain types of somatic damge. Consequently, the term tolerance dose has been not employed now-a-days but has been replaced by the term maximum permissible dose (MPO). In 1958, the I.C.R.P. postulated the following definition of MPD:

'The permissible dose for an individual has been that dose, accumulated over a long period of time of resulting from a single exposure, which in the light of present knowledge carries a negligible probability of serve somatic or genetic injuries; furthermore, it has been such a dose that any effects that ensue more frequently get limited to those of a minor nature that would not be considered unacceptable by the exposed individual and by competent medical authorities. Any serve somatic injuries (e.g., leukemia) resulting from expsure of individuals to the permissible dose would get limited to an exceedingly small fraction of the exposed group; effects such as shortening of life span, which might be expected to take place more frequently, would be very slight and would likely be the hidden by normal biological variations. Thus permissible doses can be expected to produce effects that could be detected only by statistical methods applied to large groups.'

**Balancing the Risk**

The problem in setting MPD values has been to balance the expected benefits to get derived from the use of radiation and atomic energy against the risk from radiation exposure. If there is the absence of any definite evidence of a threshold dose for certain types of somatic damage, it must be regarded that there is some risk associated with a dose of any size. It may be very difficult to define the proper balance between risk and expected benefit and it may vary considerably from country to country, depending on social and economic factors. Thus, the MPD values given by ICRP have been only recommended values and each country must be able to accept or modify them in accordance with its own particular risk-benefit conditions.

**Recommendations of I.C.R.P.**

The Commission's recommendations have been attempting to limit somatic effects in individuals and hereditary effects in the population as whole. It therefore becomes necessary to consider not only dose to individuals but also the average dose to the population. Table 11 gives the list of the appropriate dose limits for adult exposed in the course of their work and also for members of the public. We know that some organs are more sensitive than others to the effects of radiation. Therefore, different limits have been specified for various organs.

**Table 8**

*Summary of Does Limits for Individuals*

| *Organ* | *MPD for adults evposed occupationally* | | *Members of the public* |
|---|---|---|---|
| | *Rem / year* | *Rem / 13 weeks* | *Rem / year* |
| Gonads, red bone marrow | 5 | 3 | 0.5 |
| Skin, bone, thyroid | 30 | 15 | 3.0 |
| Hands, forearms, feet and ankles | 75 | 38 | 7.5 |
| Other single organs | 15 | 8 | 1.5 |

The following points must be stressed in applying these does limits :

(i) All unnecessary dose must be avoided.

(ii) Adults exposed in the course of their work who might exceed 3/10ths of the MPD should be subject to medical supervision and personal monitoring. Such persons are often called Classified Radiation Workers.

(iii) The permissible quarterly (13 week) doses for occupational exposure have been based on one-half of the annual dose, but are rounded up. The object of the 13 week MPD's has been to limit to some extent the rate of accumulation of dose. For example, if only annual limits were specified it would be possible for a worker to receive 5 rem (whole body) on the 31 December and another 5 rem on 1 January, providing a total of 10 rem in 2 days. This is obviously undesirable and is only permitted in very special circumstances (see below).

(iv) In special circumstances it may be justifiable to allow the quarterly quota to get repeated in successive quarters provided that the total body dose accumulated dose not exceed 5 (N - 18) rem, where N denotes the age of the person in year. Thus it has been permissible under special circumstances for a person to receive in a single years dose of 12 rem to the gonads and rad bone marrow always provided that the D = 5(N - 18) rule is obeyed.

(v) Women of reproductive capacity should not get exposed at the increased whole body rate of 3 rem/13 weeks, but should get limited to 12.5 rem/13 weeks.

(vi) The suggested limits for members of the public should be in all cases a factor of 1/10 of the equivalent limits for adults exposed in the course of their work.

(vii) In the case of uniform whole-body irradiation, the gonads and rad bone marrow are regarded to be the critical organs, so for uniform irradiation the whole body dose limits have been the same as those for the gonads and red bone-marrow.

**Planned Special Exposures**

In very special circumstances, it has been permissible to receive doses of up to twice the annual limit provided that the 5(N-18) rem limit does not get exceeded and the quarterly limit does not get exceeded in the last twelve months. Planned special exposures should not be allowed to a worker who has previously received abnormal exposures in excess of five times the permissible annual dose, or to women of reproductive age. It is generally felt that planned special exposures should only be allowed under very unusual circumstances where no alternative solution has been possible.

**Abnormal Exposures in Emergencies or Accidents**

It is obvious that it is not possible to specify dose limit for accidental exposures. However, I.C.R.P. have recognized the need to protect the livelihood of a worker involved in such an accident. They recommend that no special limitations should be kept on his future radiation dose following a 'one-in-a-lifetime' accidental exposure of less than 25 rem.

It is important to remember that when a worker has used up his once-in-a-life-time special exposure allowance he severely gets limited as far as future radiation work goes. After such an exposure, he must be used on work which has an extremely low probability of leading to another high dose.

In an emergency, male valunteers may get large doses for the purposes of rescue or of saving valuable installations. It has been difficult to specify limits to cover such an event as each situation will be unique. It has been often recommended that a whole body dose limit to 10 rem should be applied if saving of life is not involved. When rescue operations are to be carried out it may not be possible to work to specified dose limits. Each situation has to be assessed by the competent person or persons, and a decision reached on the basis of this assessment.

It has been probable that whole-body doses of up to 100 rem could be acceptable in the saving of a life. If the operation would need doses much in excess of this level, then the risks and possible result of the operation would be judged very carefully. One important point would be the accuracy of the information regarding the likely dose-rates in the accident area and a second, the condition of the casualties and their likelihood of survival. For example, if the dose-rate estimates were low by a factor of two, a rescuer might unwittingly receive a sufficiently large dose to produce severe early somatic effects.

**Exposure of Population**

The main contributors to the exposure of the general population have been natural radiation, medical uses of radiation, the increasing use of radioactive substances in industry and the release of readioactive material to the environment. Two methods have been employed to keep this exposure as low as possible, such as the reduction of individual doses and the limitation of the number of persons exposed.

The genetic dose to the population refers to the average received by each person from conception to the mean age of child-bearing, which has been taken as 30 years. A provisional limit of 5 rem per generation, in addition to the dose from medical procedures and natural background, has been recommended.

When estimating the genetic dose to the population at large the annual 'genetically significant' dose has to be multiplied by the mean age of child bearing, i.e., 30 years. The annual 'genetically significant' dose to a population may be defined by I.C.R.P. as the average of the individual gonad doses, weighted in each individual for the expected number of children conceived subsequent to the exposure.

In its publication 1, I.C.R.P. had included an Illustrative Apportionment, which showed how national authorities might apportion the various contributions to the genetic dose. In a later report, however, the Commission has not included an Illustrative Apportionment because of the danger that the values given in such an Apportionment might be regarded as recommended values for genetic exposure. It simply provides an indication of the magnitude of the main contributors to the genetic dose, which may be listed as follows:

(i) *Occupational exposure.* The latest available data indicate that the genetic dose from occupational exposure has been at present probably less than 0.01 rem. The increasing use of atomic energy and other sources of ionizing radiation will give rise to an increase in the number of occupationally exposed individuals. However, if the individual doses received by these occupationally exposed workers continue at the present level, the genetic dose from all occupational exposure would likely to remain well below 1 rem.

(ii) *Miscellaneous exposure.* Extensive surveys reveal that the genetic dose to the current generation may lie between 0.2 and 2 rem with an average value of about 1 rem.

It is evident that medical exposure will contribute about 1 rem of the genetic dose to the current generation. The contribution from all other man-made sources has been at present probably less than 0.2 rem. The total genetic dose has been thus at present considerably less than the recommended limit of 5 rem and it seems unlikely that this limit will get approached in the near future.

## Detection of Ionising Radiation and its Measurement

### *General Principles*

The human body fails to sense ionizing radiation. It has been probably responsible for much of the general apprehension about this type of hazard. Reliance must be kept on detection devices which have been based on the physical or chemical effects of radiation. These effects include:

(a) ionization in gases,

(b) ionization and excitation in certain solids,

(c) changes in chemical systems, and

(d) activation by neutrons.

The majority of monitoring instruments employ detectors based on ionization of a gas. Certain classes of crystalline solids show increases in electrical conductivity and effects attributable to excitation, including scintillation, thermoluminescence and the photographic effect. Detection system are known in which chemical changes could be measured but these have been rather insensitive. A method which may be applied to neutron detection depends on the activation caused by neutron reactions.

Radiation could be measured in either the differential or the integral mode. The former is able to record the number of individual ionising events, differentiating (or distinguishing) between one event another. While the latter measures their total cumulative effect integrating the individual effects from the separate events.

The measurement of radiation in radiotrace work, commonly called counting, has been concerned with the determination of activity in terms of disintegration rate, and has been generally a differential process. By contrast, the measurement of radiation has been in terms of the dissipation and absorption of energy in biological systems and has been almost an integrating process.

**Radiation Dosimetry**

*Introduction*

A number of instruments have been developed to measure the amount of radiation energy absorbed in a substance. They are called dosimeters or dosemeters. Most of these instruments are based on the radiation detection and measurement methods described earlier. There are two broad categories of these methods and instruments based on them. Absolute methods involve the determination of radiation intensity, or dose, from physical measurement of say, charge carried by a beam of charged particles of known energy or energy absorbed in a medium or the ionization produced in a gas. For routine use, however, it is more convenient to use secondary dosmeters which have been calibrated in terms of absolute instruments. Chemical dosimeters, film badge dosimeters, etc., are included in the latter category.

Radiation dosimeters are basically radiation detectors with the additional requirement that they are compact, rugged, and easy to handle. Both direct and indirect effects if radiations like ionization of a gas, heat involved in a calorimeter, chemical effects in liquids, solutions and emulsions, thermoluminescence, thermal currents and florescence produced by electrons generated in a solid medium by radiation, have been used in various types of instruments.

Most of the instruments used for personnel monitoring are intergrating type. The absorbed radiations energy is accumulated over a certain period of time. Then the effect of absorbed radiation is a measure of the dose received by a worker.

**1. Thermoluminescent (TL) Dosimeters**

Thermo-luminescent (TL) dosimeters have been currently among the most reliable and accurate instrument which are useful in a very wide range of does from 1 μGy (0.1 m rad) to 10 Gy (1 K rad). A crystalline material such as calcium fluoride ($CaF^2$) or lithium floride (LiF) or a similar compound in powder form is incorporated on a plastic material like teflon. Irradiation of the material produces electron trapping at the crystal lattice sites at both shallow and deep traps. On heating the material the electrons are released. They produce luminescence on de-excitation. A photo-multiplier measures the amount of light produced. The light output has been proportional to the received dose. A typical TL Dosimeter is in the form of a pen (or ring) in which the sensitive material (LiF crystal rod) and filament are placed in a small vacuum tube. The filament is heated when the tube is plugged in a voltage supply. A number of manufacturers market complete TLD systems for the preparation of TLD badges and arrangement for their processing after their exposure to radiation. Finger monitors in the form of rings of TLD material for monitoring the additional dose received by the fingers and hands are also available.

To charge a pocket chamber or desimeter, a charger containing a battery or rectified power supply is used to apply voltage to the chamber electrodes. The voltage is adjusted until the shadow of the fiber falls at 0 on an illuminated scale. After exposure to radiation, a pocket chamber must be reinserted into the charger (termed the charger-reader) to determine the deflection of the fibre upscale. The pocket diosimeter contains its own scale, and the deflection of the fibre upscale may be tetermined by holding the dosimeter toward light.

When both neutrons and gamma rays are to be measured the TL dosimeters with discs of LiF prepared from separated $^6$Li and $^7$Li isotopes are used. $^6$Li has a high cross section for $^6$Li $(n, \alpha)^3$H reaction with thermal neutrons whereas $^7$Li is relatively insensitive to neutrons. Both the isotopes are however equally sensitive to gamma rays. Filters are used for discriminating between beta and gamma radiations. By using lead filters the response of the device becomes energy independent for X-and gamma rays in an energy range of 20 keV to 20MeV.

The TL dosimeters have certain memory component. When they are heated for a short time only the shallow traps are emptied of electrons to produce luminescence. On further heating to a higher temperature the electron at deep traps produce light output. In this manner it is possible to make a second measurement of the same dose at a later time.

A TL dosimeter is used over and over again. This is done first by erasing the memory of an earlier dose by heating the dosimeter to a high temperature. Then it is ready for making fresh measurement of dose.

**2. Thermo-current (TC) Dosimeters**

An improvement-cum simplification over the thermo-luminescent (TL) dosimeters has been achieved by developing thermo-current dosimeters. The principle of this device has been similar to that of TL dosimeters. Radiation produced electrons are trapped at impurity levels in a thin crystal of synthetic saphire which is 10 mm in diameter and 1 mm or less in thickness in a typical device. The trapped electrons get liberated on heating to produce current in a circuit formed by two electrodes deposited on the crystal surfaces. Then the peak value of current or the integrated current is a measure of absorbed radiation dose. This device eliminates the use of photo-multiplier tube. Further it has been more convenient to measure small currents than feeble light output. Therefore the TC dosimeters are likely to replace the TL dosimeters. The latter have already replaced the film badge dosimeters to a large extent.

**3. Film Dosimeters**

Photographic emulsions have played vital role as integrating detectors for directly and indirectly ionizing radiations. Photographic films have long been in use in dosimetry, especially with X-rays. Under controlled processing and good densitometric measurements film dosimeters have given generally reliable results. A typical film badge is in the form of a cassette 4 cm by 5 cm in size with several windows and a number of filters. A dental X-ray film or a special nuclear emulsion film is mounted in the cassette. Filters of different materials like plastic,lead-cadmium, lead-tin, dural and indium absorber, are used to discriminate among the various type of radiations with different energies.

An exposed film and a reference film are processed together. The radiation exposure time is usually one week but it can be less if an accidental over-exposure has taken place. Measurements of the densities of the blackenings of exposed and reference films and their comparison with suitably calibrated standards with known radiation doses give the radiation dose to which the worker has beeen exposed.

The normal range of doses measured with these devices has been 500μ Gy to 200μ Gy (0.5 to 20 rads). These limits can be extended by using different types of films. Nuclear emulsions are used for measurement of dose from fast netrins and high energy particles produced by accelerating machines. Special boron loade films are used for the thermal neutrons.

Although the film badge dosimeters have been in use for several decades and are still being used in several establishments as routime personnel monitors, they are being replaced by TL and TC dosimeters.

**4. Direct Reading Ionization Chambers**

The principle of ionization of a gaseous medium by radiations, particularly the X-and gamma rays, has been used in condensor or pocket non-chambers. A gas contained between two electrodes kept at a potential difference of 100 to 150 V is ionized by the incident radiation. The ionization is a measure of energy and intensity of incoming radiation. While positive ions move to the methods, which is generally the case of the instrument, the electrons are collected at the central anode. The amount through which the anode potential has dropped in a given times is a measure of the gas ionization and hence the exposure from incident radiation during that interval. In a direct reading instrument, like a pocket type or a pen type ion chamber, the movement of metallized quartz fibre across a calibrated scale indicates the total dose received. For more accurate measurement auxiliary circuitry is used with other versions.

These instruments are generally used for the measurement of daily dose received by the workers in contrast with the film badges which measure weekly doses. The capacitor in charged to a full voltage at the beginning. As the exposure builds up the capacitor gets discharged. The reading on the scale, calibrated in terms of roentgen, rads or Gy gives the dose received at the end of a working day or an operation. These instruments have ranges from 200 Gy to 10 Gy (20 mrad to 1000 rad). Usually, instruments with 1000 to 2000μ Gy (100 to 200 mrad) full scale ranges are used in the moral laboratory work. Chambers lined with boron are used in work with slow neutrons.

*Auritsen Electroscope* : This device is similar to the pocket dosimeter, except that it is bulkier (the chamber is 2-3 in, in diameter and about 3 in, long) and is used primarily for measurement of radiation from radioactive samples. In the instrument, the deflection of the quartz fibre is viewed through the microscope at one end of the chamber.

The Landsverk electroscope or analysis unit is somewhat similar to the Lauritsen electroscope, except that the sample may be inserted and measure within the chamber.

The size of the pulse produced by single ionising particle may be easily calculated as follows. About 30 eV are needed for an alpha to form an ion pair in a gas. A high energy alpha of 60 MeV will produce about 2 x $0^5$ ion pairs. Now one electron carries a charge of 1.6 x $10^{19}$C, so that the arrival of this seemingly large number of electrons at the anode will generate a charge of only 3.2 x $10^{-14}$C. It is therefore apparent that ion currents, even with large fluxes of energetic particles, are very small. These will be even smaller with the smaller fluxes of lower energy particles that are more commonly encountered, as with the beta-emitting isotopes used in tracfer experiments. Such small currents are difficult to measure directly. In instruments other than the electroscope they are amplified before measurement.

**5. Phosphate Glass Dosimeters**

Dosimeters described above are useful for personnel monitoring. They have a narrow and limited ranges of operation. In problems connected with high doses of radiation like those in atomic blasts; reactor or accelerator flux densities; irradiation of foods, medical supplies etc., glass dosimeters are handy.

A glass dosimeter has been a small strip of a special type of glass, a few cubic cms in size, containing small amounts (5 to 10 per cent) of a phosphate with silver impurities. Usually silver phosphate in polymer form is used. The strip is coated with a thin layer of an opaque material to protect it from visible light. Ionizing radiation produces electrons which are trapped at impurity centres and stay there permanently until released by ultra-violet radiation. Then give rise to liminescene which is measured with a photo-multiplier-recorder assembly. The amount of emitted light is a measure of dose received at the dosimeter. This devices could be used for recording doses from a few Gy to several k Gy which are used in irradiation units for preservation of foods and sterilization of medical equipment.

**6. Chemical Dosimeters**

A second type of dosimeters suitable for measurement of a dose in the ranges of $10^2$ to $10^5$ Gy makes use of the chemical effects of radiation. A large number of possible variations have been suggested. One of the most popular types employs chloroform which forms hydrochloric acid on radiolysis. A dye is used as an indicator of the pH-value of solution which depends on the amount of HCI formed. In one version choloroform in aqueous solution and the indicator dye are contained in a sealed glass tube. Irradiation of solution produced a change in the colour of the dye. The dose is then obtained from the extent of change of colour.

Fricke dosimeter, which uses ferrous solution, is based on the oxidation of ferrous ions ($Fe^{2+}$) into fetric ions ($Fe^{3+}$) under irradiation. This dosimeter is used in X-ray dosimetry. An aqueous solution of ferrous ammonium sulphate (0.001 M), sodium chloride (0.001M), and sulphuric acid (0.4M) is usually employed. Radiation changes $Fe^{2+}$ ions into $Fe^{3+}$ ions. The concentration of the latter is then a measure of radiation dose. The amount of $Fe^{3+}$ ions is determined by spectrophotometric method. This dosimeter has been used to measure doses as high as $10^7$ to $10^8$ Gy per sec. Its response has been independent of the dose rate upto a dose of $10^5$ to $10^6$ Gy per sec but depends on the presence of absence or oxygen.

Ceric sulphate dosimeter gets on the reduction of ceric ions ($Ce^{4+}$) into cerous ions ($Ce^{3+}$) under irradiation. Its response is independent of the presence or absence of oxygen. The dose is measured from the yield of $Ce^{3+}$ ions. This is obtained from the difference in ceric ion concentrations before and after irradiation. Optical and analytical methods are used for measurement of $Ced^{4+}$ ion concentration. This dosimeter shows a linear response in a region of $10^3$ to $10^5$ Gy. Doses upto $10^6$ Gy could be measured with this type of dosimeter.

**7. Calorimetric Methods**

Absorption of radiation by a medium raises its temperature. If the changes in temperature could be neasured accurately then it becomespossible to determine the amount of energy deposited in the medium by the incident radiation. This forms the basis for calorimetric methods. For an ionizing radiation imparting 10 Gy of dose to a material the rise in temperature is only a few millidegrees. The success of this method therefore depends on the accuracy with which small changes in temperature could be measured.

The importance of calorimetric method lies in its being an absolute method against which the other indirect dosimetric measurements may be checked.

For an experiment, a sample of water or graphite is kept in a thermostatically controlled enclosure, insulated thermally from the environment. It is usual to follow the small temperature drifts before and after the measurement of rise in temperature due to a beam of incident radiation. As the minor drifts in temperature can not get eliminated completely therefore an extrapolation procedure is used to determine the rise in temperature. A precision of about 1 per cent in the measurement of temperature difference is unually obtained. The calorimetric method gives the rate of energy production in the medium. This gives the activity of the source if the average energy of alpha or beta particles emitted by the source is known. Energy production rates of the order of a few micro-joules per sec have been measured with an accuracy of 1 per cent.

It becomes necessary to ensure that radio-chemical reactions do not modify the changes in temperature. This is usually carried out by pre-irradiation of the system. In that case any chances of further chemical changes could get reduced.

**8. Radiation Monitors**

A different type of instrument is required when monitoring the area of actual work or some operation where radio-active source are being used. A variety of instruments including ionization chambers, GM-counters, and scientillation counters with rate-meter dispkays, an available commercially. They are usually known as survey meters or contamination monitors and calibrated in terms of counts per sec or counts per min or m Gy (or rad) per hour. Both mains operated and portable varieties are commonly available for the laboratories.

A survey meter with ionization chamber detector, dc amplifier, and calibrated meter has ranges from 200 μGy per hour to 30 m Gy per hour. A removable cap, which works both as a protection for thin window as well as a beta particle shield, enables the monitoring of both beta-gamma and gamma activities.

A portable Gm-counter Survey Meter with cylindrical type of countries is more sensitive instrument with ranges from 2 μGy $h^{-1}$ to 200 μGy $h^{-1}$ (0.2 m rad $h^{-1}$ to 200 m rad $h^{-1}$). Here again a thin shield which covers the tube allows the instrument to detect only gamma rays above 100 keV energy while with the shield removed it is sensitive to both beta (above 0.2 MeV) and gamma rays. Thin window GM tubes have been very useful probes for beta particle monitoring. Soft beta rays from $^{14}C$ and $^{38}S$ need extremely thin windows.

A Scintillomwter using Nal (TI) Photomultiplier rate-meter system has still better sensitivity for gamma rays. It measures radiations does as low as 0.01. Gy per hour. Thin Nal (TI)crystal (1 mm thick) scintillation probes are available for use in 5 to 50 leV region. They are particularly useful for low energy photons from $^{325}I$.

Contamination monitors with warning systems also form part of radiation protection equipment. An instrument fitted with a small speaker or some other relay operated warning system becomes operative when the radiation dose exceeds a certain limiting value.

Levels of radio-activity in air are monitored with radio-active gas monitors. Air is withdrawn through an ionization chamber. The level of radio-activity due to alpha or beta emitters in gaseous forms has been then indicated by audible or visual displays. In an another version a small vacuum cleaner is used to draw out air and dust into a filter paper. The latter could be then monitored for its radio active content.

**Nuclear Energy and the Environment**

In assessing the radiological effects of nuclear technology, the individual power plant is only "the tip of the iceberg". To enable it no operate, a nuclear power plant has to be linked with a fuel cycle that emits radioactivity at many points and so contaminates the environment.

The first stage in the nuclear industry is the uranuim mine. The ore is pulverized and subjected to a leaching process to separate the uranium. The waste from this separating process, known as tailing, contains the subsidiary products of the natural radioactive series and is deposited on dumps. In the United States alone, the volume of these tailings already amounts to some 100 million tons. Rainwater washes the soluble nuclides into the ground, into groundwater and into surface waters, whence it is absorbed by organisms in the soil and water. By the extraction of uranium from the mines, the natural radioactive substances that normally lie deep down in the earth's crust are brought to the surface and released into the biosphere.

The uranium extracted from the mines then passes into the factories for preparing fuel elements, where the fissionable uranium-235 isotope recovered from nature in small quantities is enriched. In this highly energy-consuming enrichment process further radioactive waste is formed and given off into the environment. Once the uranium has been sufficiently enriched it is fed, in the form of small tables, into fuel elements for nuclear power plants. In the nuclear power plant the real process of nuclear fission takes place, whereby a large number of mostly radioactive fission products are formed from the uranium-235. These partly gaseous radioisotopes escape to some extent in the nuclear power plant and are given off into the air and water below the legally permitted limits for radioactivity.

After about a year the fuel elements are exchanged and, after a period of intermediate storage in the nuclear power plant to permit decay of the short-period isotopes, are transported to a reprocessing plant. For reconditioning purposes the fuel cells are split by remote control, the individual roads are cut into short sections, and the uranium oxide is washed out of the sheaths with nitric acid. By this disintegration of the fuel elements the radioactive inert gases cannot be formed by nuclear fission are released. As inert gases cannot be combined by a chemical process, they are emitted from the reprocessing plant through a chimney into the atmosphere. Other radionuclides released in gaseous and liquid form in this process are kept out of the waste air and water of the plant means of filters. But as these filters are not 100 per cent absorbent, a certain amount of the radioisotopes escapes into the environment.

Overall the emission of radioactivity by reprocessing plant, related to the corresponding fuel equivalent, is higher by a factor of 100-200 than in nuclear power plants.

The uranyl nitrate solution produced in the dissociation of the fuel elements is subjected to a multistage extraction process, by means of which the uraniuim the plutonium and the fission products are separated from one another. The refined end-products, uranium and plutonium, are processed again by manufactures to provide new fuel elements.

**Biological Effects of Ionizing Radiation**

Radioactive substances have been among the most toxic materials known. The magnitude of their injurious effect has been in a range that might be said figuratively to represent a "quantum jump" in comparison to ordinary organic chemical poisons. Radium has been 25000 times more lethal than arsenic.

The biological importance of radiation were known when in 1895 Wilhelm Roentgen kept his hand between the newly discovered x-ray tube and a flourescent screen and noted that the bones cast a deeper shadow than the ffesh. The following year, the French physicist Antoine Becquerel discovered natural radio activity in a uranium compound, a discovery for which he shared the Nobel Prize with the Curies in 1903. An accident with one of the earlest known radioactive substances gave Bequerel the first indication of the biological potency of the newly found radiation when he got burnt himself with a vial of radium that he was carrying in his pocket.

The most tragic early evidence of the potency of radiation had been death of Marie Curie, although the lesson seemed to go unheeded for many years. Marie Curie with the help of her scientist husband. Pierre Curie, found that uranium are had been having a substance as yet unidentified that could be several hundred times as radioactive as the uranium with which she was working. They were successful in isolating from the ore a radioactive element which they named polomium after Maries native country Poland. But polonium accounted for only a fraction of the radioactivity, and in 1898 they identified radium, an intensely radioactive substance. Pierre, upon learning of Bacqerel's experience with radium, carried out an experiment that deliberately produced a burn on his own arm.

Marie Curie became famous. Although she was denied membership in the snobbish French Academy by one vote because she was a woman, she was able to share the 1903 Nobel Prize with her husband and Bacquerel. In 1911, she again received the Nobel Prize, this time for the discovery of two new elements. Marie Curie died of leukemia, a cancer of the formative white blood cells of the body, almost certainly brought on by the high levels of radiation to which she was exposed during her work. Irene Joliot-Curie, the scientist daughter of the husband and wife team, also died, of leukemia apparently from radiation exposure.

When x-rays were discovered, their usefulness in medical diagnosis and other applications was quickly recognized. However, the dangerous nature of x-radiation and its potential for serious, even fatal, injury were largely unknown. Indiscriminate use of x-ray machines brought about numerous cases of injury and illness. Miners suffered toxic effects from radium and thorium and the careless use of radioactive materials, most notably radium, made serious ailments and premature deaths among workers. The potency of x-rays and radionuclides has been so great that the enormously expanded use of ionizing radiation in its many forms in research, medicine and industry is potentially hazardous people working in areas of high radiation and in the environment at large.

The interaction of ionizing radiation with the human body, arising either form external sources outside the body or from internal contamination of the body by radioactive substances, give rise to biological effects which may later show up a clinical symptoms. The nature and severity of these symptoms and the rate at which they appear depends on the amount of radiation absorbed and the rate at which it has been received. Radiations injuries could be divided into two classes, somatic effects in which the damage appears in the irradiated person himself and genetic effects which arise only in the offspring of the irradiated person as a result of radiation damage to germ cells in the reproductive organs-the gonads.

The radiation released upon the disintegration of radioactive atomic nuclei takes the form of particulate or wave radiation of very high energy. When this radiation impinges on living cells, various interactions can result. The most important of these is the ionization of atoms. In the electron sheath of an atom, a certain number of negatively charged electrons are present which correspond to a like number of positively charged protons in the atomic nucleus. Consequently an atom is normally electrically neutral in its outward effects. If radioactivity, such as a negatively charged electron of beta particle radiation, bombards a neutral atom, it may be that owing to the high energy of the radiation one or more electrons will be ejected from the electron sheath of the atom. The atom then becomes an ion a positively charged (electrically) atomic residue. Owing to their electrical attractive power, newly formed ions have a capacity for strong chemical reaction. If this takes place in dead matter, in the air for instance, it is of no importance. An ionized nitrogen atom, for example, can combine with ionized oxygen atoms; and ionized oxygen atoms can combine three at a time to form one ozone molecule. But if this process of ionization of atoms takes place in living tissue, in certain circumstance a variety of damage may result.

It is now known that the plan for the constitution of a living organisms is contained in a long filamentary structure, in which, on the basis of a percise arrangement of small molecules, information regarding heritable characterists is contained. This is deoxyribonucleic acid (DNA) and it consists of a basic structure of sugar and phosphate molecules to which four different basic molecules are attached in a particular sequence. This sequence is what is known as the genetic code, which represents the heritable information. In the case of each cell division the DNA is also divided and restructured on identical lines, so that the heritable information is fully retained in each cell in the case both of the restructuring of body cells. If radioactive rays now impinge on a DNA molecule, there is a danger that, with the ionization of atoms and the chemical reaction ensuring therefrom, a variation may occur in the basic arrangement and so in the heritable information. Such a variation in the heritable information is known as a mutation. Almost all the mutations are negative, it is estimated that over 99.99% of the mutations change the organism's characteristics negatively. If the mutation occurs in germ cells that is in male sperm cells or in female egg cells-this can result, when a new organisms is formed, in severe damage, such as deformities, stillbirths, enzyme defects amd other metabolic disorders. In body cells, otherwise known as somatic cells, a mutation can cause cancerous degeneration.

Ithas to be assumed that not tolerance limits exist as regards this type of damage that is to say, that the very smallest quantity of radioactivity is capable of causing such damage. High radiation exposure can touch off a chain reaction via ionization of atoms, chemical reactions, toxic molecules reaction and poisoning of the entire organism, leading to immediate danger in the form of fever, nausea, vomiting, loss of hair and death from radiation.

**Types of Biological Effects**

Ionizing radiation is different from ordinary environmental contaminants in several ways. One of the most important is that some of the radioactive elements continue to emit radiation for extremely long periods of time—in some cases for thousands of years comprising a special kind of hazard to man and to the environment. The increasing quantity of radioactive waste products poses problems that are unique and require greater knowledge than is now available concerning both physical and biological aspects of the long-term effects.

**The Interaction of Radiation with Cells**

The basic difference between nuclear radiations and the more commonly encountered radiations such as heat and light is that the former have sufficient energy to cause ionization. In water, of which cells are largely composed, ionization can lead to molecular changes and to the formation to the chromosome material. The damage takes the form of changes in the construction and the function of the cell. In the human body, these changes could be able to manifest themselves as clinical symptoms such as radiation sickness, cataracts or, in the longer term, cancer.

The process leading to radiation damage could be conveniently divided into four stages as follows :

(i) *The initial physical stage*, lasting only a minute fraction ($\sim 10^{-16}$) of a second in which energy gets deposited in the cell and causes ionization. In water the process may be written as follows :

$$H_2O \xrightarrow{\text{radiation}} H_2O^+ + e^-$$

where $H_2O^+$ is the positive ion and $e^-$ the negative ion.

(ii) *The physico- chemical stage*, lasting about $10^{-6}$ seconds in which the ion interact with other water molecules resulting in a number of new products. For example, the positive ion dissociates,

$$H_2O^+ \longrightarrow H^+ + OH$$

the nagative ion, that the electron, attaches to a neutral water molecules which then dissociates

$$H^2O + e^- \longrightarrow H_2O^-$$

$$H_2O^- \longrightarrow H + OH^-$$

Thus, the products of the reactions have been $H^-$ $OH^-$ H and OH. The two former ions,which are present to quite a larger extent in ordinary water, take no part in subsquent reactions. The other two products H and OH, are called free radicals, that is they have an unpaired electron and have been chemically highly reactive. Another reaction product has been hydrogen peroxide $H_2O_2$ which is a strong oxidizing agent.

$$OH + OH \longrightarrow H_2O_2$$

(iii) *The chemical stage,* lasting a few seconds in which the reaction products interest with the important organic molecules of the cell. The free radicals and oxidizing agents may attack the complex molecules which form the chromosomes. They may, for example, attach themselves to a molecules or cause links in long chain molecules to be broken.

(iv) *The biological stage* in which the time scale varies from tens of minutes to tens of years depending on the particular symptoms. The chemical changes described above can affect an individual cell in a number of ways. For example, they may result in :

(a) the early death of the cell,

(b) the prevention or delay of cell division, or

(c) a permanent modification which is passed on to daughter cells.

The effects of radiation on the human body have been the result of damage to the individual cells. These effects may be conveniently divided into two classes, namely somatic and genetic. The somatic effects arise from damage to the ordinary cells of the body and affect only the irradiated person. The genetic effects, on the other hand, are due to damage to the cells in the reproductive organs, the gonads. The important difference is that, in this case, the damage may be passed on to the person's children and subsequently to later generations.

When ionizing radiation happen to pass through body tissues, it is able to split the molecules of the tissues into ions or free radicals, as in the examples of water molecules. These fragments may later combine to form new chemical compounds (such as)$H_2$ from H+H or $H_2O_2$ (hydrogen peroxide) from OH+OH) or free radicals (such as $HO_2$ from H H+$O_2$). Ionizing radiation can thus split molecules into useless or reactive fragments and allow the formation of other reactive compounds. These effects can be classified into two groups:

1. *Direct effects.* Fragmentation of biologically important molecules, such as DNA molecules in the cell nucleus.

2. *Indirect effects.* Fragmentation of biologically less vital molecules (such as water) with the formation or reactive ions or free radicals that can later affect more important molecules and impair their usefulness.

The damage caused by readiation has been found to depend on the amount of energy that gets lost by the radiation and made available for the ionization for matter. The amount of energy lost by the radiation per unit distance travelled (which is not necessarily in a straight line-beta rays, for example, are easily detected by matter) is termed as the radiation increases as the energy of the radiation decreases, in part because its speed will be less and the traversal of a unit distance will require a longer period of time.

Low-LET radiation does most of its tissue damage through indirect actions, while high-LET radiation does most of its damage through direct action. A particle entering tissue with high energy will have low LET initially but as its energy decreases, its LET increases. Such a particle could be used in treating a deep-seated tumor if the particle's range is just right for it to come to a stop near the tumor.

The site in a living cell most vulmerable to ionizing radiation has been its nucleus, where most of its genetic material (DNA) has been located. If ionizing radiations are able to damage the molecules of this chemical, the chemical of genes, the cell may be made reproductively dead (not able to divide), or the cell may produce flawed "daughter" cells when its does divide. If the radiations is having affected germ plasm, the flaws may eventually give rise to permanent genetic damage to offspring; if the damage has been in somatic tissue,they may give rise to tumors or cancer. Radiations in to bone marrow, for example can give rise to leukemia. Radiations into the pelvic region of a pregnant woman may cause damage to the fetus. X-rays and gamma rays penetrate intact skin ; alpha rays and beta rays cannot.

**Table 9**

*Average LET Values and Tissue Penetration Distances for Electrons, Protons, and Alpha Particles.*

*1 keV = 1000eV (Electron Volts) = $1.6 \times 10^{-16}$ Joules.*

*$1\ \mu m = 10^{-6} m$*

| *Particle* | *Energy (keV)* | *LET (keV / μm)* | *Tissues penetration distances (μm)* |
|---|---|---|---|
| Electron | 1 | 12.3 | 0.01 |
| | 10 | 2.3 | 1 |

| | | | |
|---|---|---|---|
| | 100 | 0.42 | 180 |
| | 1000 | 0.25 | 5000 |
| Proton | 100 | 90 | 3 |
| | 2000 | 16 | 80 |
| | 5000 | 8 | 350 |
| | 10000 | 4 | 1400 |
| | 200,000 | 0.7 | 300,000 |
| Alpha | 100 | 260 | 1 |
| | 5000 | 95 | 35 |
| | 200000 | 5 | 20000 |

Radiations enter the body on airborne, radioactive dusts and gases, or in food. The death rate from lung cancer in the infamous uranium mines of Joachimsthal, Bohemia was 30 times normal in the 1930s. Radon, a radioactive decay product of uranium and radium, has been a chemically inert but radioactive gas that did not dissipate readily in the mines, which have been worked for lead, cobalt, and arsenic, and the frequency of "lung diseases" among the miners was noted more than 400 years ago.

As radiation damage in a cell has been most serious when it affects the generic chemicals that are critical in cell division, the symptoms of such damage appear earliest among tissues whose cells divide most frequently. Tissues forming the walls of the stomach and intestines and the tissue that makes white blood corpuscles in bones have such cells. It has been not surprising, therefore, that early symptoms of radiation damage involve the stomach (nausea and vomiting) and intestines (diarrhea), as well as a drop in the supply of white blood cells. Strontium 90 (half-life 27.7 years) has been dangerous because it has been a bone-seeker,being chemically similar to calcium, a component of bone. Cesium 137 (half life, 26, 6 years) is chemically similar to sodium, an element present as the sodium ion in ordinary table salt. Ions of cesium 137-are easily distributed throughout the entire body, including the gonads.

From a biological point of view, radiation effects are also classified as somatic or genetic. Somatic effects have beem the effects on the body itself and have been of direct concern to the person exposed to the radiation. Genetic effects have been those imvolving mutations of the chromosomes or genes in sex cells; they pose a potential hazard to the descendants of the exposed person and are of concern to the whole society. Genetic effects have been of grave concern only in persons who have been not yet past their reproductive years.

We are lacking a basic understanding of the biological action of ionizing radiations despite a great deal progress in research. Most experience have beem carried out using laboratory animals, and comparison with humans has been difficult to make. Some data could be accumulated over the years from human exposures to large radiation doses obtained accidentally or from nuclear explosions. Table 11 lists the approximate short-term effects that might be experienced for whole body adiation exposures over a short period of time. A whole-body exposure of 1 rad would mean an average absorption of 100 ergs/s over the whole body. Table 10 refers to immediate effects, and there is no guarantee that recovery might not be followed many years later by other effects-such as greater incidence of cancer than that occurring in unexposed persons. As different persons differ in their sensitivity to radiation, as in their sensitivity to infections and numerous other things, the doses corresponding to different effects will vary widely from person to person.

**Table 10**
*Estimated Short-term Effects of Single-dose, Whole-body Radiation Exposures in Man*

| | |
|---|---|
| Less than 25 rads | No observable effect |
| About 25 rads | Threshold level for detectable effect |
| About 50 rads | Slight temporary blood changes |
| About 100 rads | Nausea, fatigue, vomiting |
| 200 to 250 rads | Fatality possible, though recovery is more likely |
| About 500 rads | Perhaps half the victimes would die |
| About 1000 rads | All the voctims would die. |

Exposure to a few hundred rads give rise to acute radiation illness : nausea, fatigue, and vomiting within a few hours and for a day or two, a decrease in red and white blood cells and blood plate-lets for a few weeks, then anemia, susceptibility to bacterial infection and hemorrhaging for some period of time, followed often by death. If the victim is able to survive, he or she will still have a greater-than average chance of developing leukemia (especially in the first few years after exposure), other forms of cancer, cardiovascular disorders, and eye cataracts.

Of particular interest in studies of radiations effects are the continuing research finding of the medical histories of thousands of survivors (and their children) of the Hiroshima and Nagasaki atom bombs of 1945.

Atom bomb survivors showed higher than average leukemia rates,which packed in 1951-6 years after exposure- but which were still higher than usual in 1966. Mortality rates for persons within 1200 m of the hypocenter in the two towns, after excluding the effects of leukemia were 15% higher than in unexposed Japanese during the decade 1950 to 1960-a statistically significant increase. No effect had been found on mortality of children conceived to survivors after the exposure, however. Very high rates of chromosomal abnormalities are found among survivors and their children who were in utero at the time of the bombings but not in children conceived afterward.

Despite frequent criticism of these studies in Japan, these studies are being continued and are playing an important role in defining the hazards of radiation. They are especially important now when increased cancer rates or other effects might be appearing after a long period of latency.

The injurious effects of radiations have been seen mainly in two forms. One is direct damage to the tissues ans organs of the exposed organism called somatic effects, from the Greek soma, meaning "body". The other is the indirect damage to subsequent generations through the disruptive action of radiation on the reproductive tissues of the parent, called genetic effects, from the Greel word genesis, meaning "descent". Genetic effects are caused by the action of radiation on the molecular or physical structure of the chromosomes.Even in the production of somatic injury, the focal point of the ionizing radiation may be its disruptive action on the DNA make up of the chromosomes of the somatic cells. Rememnber that DNA is the vital substance that has the dual function of guiding the somatic cellular activities of the individual organism as well as prepetuating racial characteristics in the progeny.

**Somatic Effects**

The first evidence of the damaging effects of radiation came from unfortunate experiences of people who, because of their occupations, either used or came in contact with sources of high-intensity radiation and suffered chronic exposure. Radiologists, uranium mine workmen, and painters of radium dials were among those who suffered disastrous effects. Patients who received large doses of X-rays and those who given natural radionuclides such as radium and thorium for therapeutic purposes were also afflicted. Additional evidence of the degree and kind of damage from radiation was forthcoming from studies of the Nagasaki and Hiroshima survivors, from the unforeseen exposure of the Marshall Island people following a weapons test in 1954, and from accidental exposures among laboratory and industrial workers. Finally, laboratory animal experimentation has been a rich source of information on the effects of ionizing radiation, dating as far back as 1927, when Hermann Muller demonstrated that x-rays can cause drastic changes in fruit flies. Many other investigators have since then shown damaging effects to animals and plants.

1. *High Radiation Exposure.* A lethal dose of penetrating ionizing radiation to humans causes an acute form of radiation sickness. Within a few hours the patient becomes nauseated and experiences vomiting and fatigue. In the typical radiation syndromes the symptoms last a day or two after which the patient feels fairly normal for a time. However, his blood is not up to par. There is a gradual decrease in the number of red and white blood cells over a period of several weeks until it is obvious that anemia has developed. The blood has lost much of its capacity to carry oxygen and the patient feels weak. Because of

the drop in white cells, he is more susceptible to infection. There is also a decrease in thenumber of blood platelets, essential for clotting. As a result, the patient begins to bleed in various parts of the body. There is hermorrhaging from the nose, gums, and intestimes. Blood clots blotch the skin and mucous membranes. If there are wounds, they heal poorly. The victim declines in vitality until he dies from anemia, infection, and hemorrhage.

Individuals that do not die within a few weeks are still not in the clear. There occurs a sharp decrease in life expectancy. Typically in mammals, such as mice, there is an interval of low mortality with a peak at about 3 months following exposure, after which the death rate increases until 9 months after exposure. Delayed deaths may occur for several months thereafter is the result of progressive deterioration from the acute effects of the exposure. During pregnancy, the fetus is more susceptible to radiation injury than adults, the effects depending on the number of days from conception.

**Table 10**

*Estimated Doses for Varying Degrees of Injury to Man*

| *Accumulated dose (R) or dose rate* | *Period of time* | *Effects* |
|---|---|---|
| 500 R/day | 2 days | Mortality close to 100% |
| 100/ Rday | Until death | Mean survival time approximately fifteen days |
| 60/R/day | 10 days | Morbidify and mortality high with crippling disabilities. |

| | | |
|---|---|---|
| 30R/day | 10 days | Disability, moderate |
| 0.5 R/day | Many months | No large scale drop in life span |

Parts of the body differ in their sensitivity to radiation injury. The most sensitive tissues from acute dosages are the intestines, the lymph nodes, and the organs responsible for regenerating blood cells-the spleen and bone marrow. The intestinal wall partially breaks down and becomes inoperative in its normal function as the main barrier to infection from the bacteria that inhabit the gut. The reduced number of leucocytes (white blood cells) that normally attack invading microorganisms are unable to cope with bacterial attack, and the body succumbs to overwhelming infection aggravated by the failure of the blood to clot because there are fewer platelets. There is also partial asphyxiation from anemia caused by the disappearance of the red blood cells.

A further effect of ionizing radiation is that it destroys the body's immune response. One of the body's defenses against extraneous substances, especially invading organisms, is a biochemical defense consisting of the rapid production of antibodies that neutralize the intruders. Thus, in addition to lowering resistance by the destruction of white blood cells, radiation impairs the ability of the body to combat infection with antibodies, leaving it vulnerable to further invading microorganisms.

2. *Fractional Dose.* The biological effect of radiation sometimes gets decreased when the total radiation dose gets divided into several exposure periods over a period of time. For example, the susceptibility of patients whose radiation had been fractionated over an 8-day period had been significantly less than those who received the same

amount of radiation in a single dose. In a typical experiment using fractioned exposures, twice as much radiation had been needed to produce a characteristic response.

3. *Low-penetrating Radiation.* The effects of the low-penetrating types of particulate radiation have been typically less severe than from penetrating radiation. This was demonstrated by a mishap that followed a text explosion at Bikini Atoll in 1954. Radioactive material was thrown up from a 15-megaton thermonuclear device. Unpredicted winds at high altitudes swept the nuclear debris off its expected course and deposited radioactive material on Rongelap, an inhabited island about 100 miles east of Bikini. Radioactive debris also fell on a Japanese fishing vessel, the Lucky Dragon, that was operating in the area.

The natives of Rongelap got suffered exposure that required medical attention and prolonged observation. There were evacuated from the atoll and it was not safe enough for them to get returned until 1957. Thc Rongelap people received a heavy dose of beta radiation from radioactive material that settled on their skins. For a few days they got bothered with considerable itching and burning but there was no erythema (redness of the skin). Then it about 3 weeks, patches of darkened skin and raised areas got appeared, especially on the scalp, and there occurred some loss of hair. Within a few months the injuries completely healed. With highly penetrating radiation like photons of x-rays, much less exposure would have been needed to cause serious injury.

4. *Delayed somatle Effects.* If the animal or human patient survives the radiation sickness from acute exposure, the possibility of illness or death

remains from one or more of several profound effects on the cells and tissues. The foremost delayed somatic effect is the increased chance that cells will become cancerous, but he appearance of cancer may be delayed for months or years. The type of cancer most commonly associated with radiation, and the first to appear, is leukemia, a cancer of the blood-forming tissues in which the white cells increase in number and become maligant. Leukemia may show up about a year after irradiatis, or even later. The chance of eventually developing one of the many other forms of radiation-induced cancers is estimated to be about five times that of leukemia.

There is also the increased probability that the victim will be afficted later in life with anemia, various cardiovascular disorders, or eye cataracts, as well as premature aging and reduced longevity. If the organism is young, he will be stunting or other impairment in growth. Fertility is reduced or temporasily impaired. Exposure of the female ovaries to a dose of 300 rads or the male testes to 1,000 rads brings about permanent sterility.

Delayed effects showed up among the residents of Rongelap Island, which received fallout from the Bikini test of 1954. The sixty-four residents of Rongelap got exposed to both gamma and beta radiation. The dose was not lethal but caused nausea and vomiting, and there was a reduction in blood cell elements that lasted for several months. Besides the skin exposure that resulted in "beta burns" and loss of hair, there was an intake of radioactive iodine with contaminated food and water. The iodine was deposited in the thyroid glands and this proved to be the most serious aspect of the exposure. Healing of the skin and regrowth of hair, as well as nearly normal recovery of the formed blood elements, were complete by the end of the first year. The effects on the thyroid glands were longer lasting.

Observers noted abnormalities of the thyroid glands ten to fifteen years after exposure. It has been highly probable that the effects had been the result of deposition of radioactive iodine in the thyroid at the time of fallout. Ninety per cent of the thyroid abnormalities had been in children who were under ten years of age at the time of exposure. Some of the children were retarded in growth and development, an effect that got related to the thyroid injury,

5. *Life Span and Aging.* Experimental work on animals reveals that radiation shortens the life expectancy of rats, mice, dogs and other animals. The causes of reduced life span are not confined to cancer but extend generally over the normal spectrum of diseases. From such data, it was estimated that at low levels of radiation exposure, the life expectancy of humans may be reduced by as much as 5 days per roentgen. This implies that, during a normal human life span, an average exposure to man-made radiation of six rem (or roentgen) would cut the life span by about 1 month. The effect is usually proportional to exposure, so the life-shortening effects of higher exposure rates, such as from excessive diagnostic or therapeutic x-rays, would be proportionately greater.

Other delayed somatic effects opf radiation have been cataracts, sterility or impairment of reproduction depending on exposure rate, and accelerated aging. These have all been demonstrated in laboratory animals. The cancer-causing effects of radiation have been extensively documented for both laboratory animals and humans. Some radiation workers believe that diagnostic radiation of pregnant women may be able to increase the risk of cancer to the child, although experiments with fetal mice show them to be surprisingly resistant to radiation before birth.

*Safety of Nuclear Power Plants*

Heightened concern about the risks of operating nuclear power plants followed a reactor accident in 1979 at the Three Mile Island power plant near Harrisburg, Pennsylvania. A series of equipment failures, misleading instrument readings and human error resulted in an overheated reactor core and prevented routine emergency systems from operating properly. Small amounts of radioactive gases were released from the plant during the crisis. Estimates varied of the health risks associated with that escaped radiation; the Department of Health, Education and Welfare predicted only 1 excess cancer death among the 2 million people living within 50 miles of the plant.

The crisis reversed earlier confidence that major reactor accidents are unlikely. The overheated reactor core might have led to further overheating and to a meltdown, the term for an uncontrolled nuclear reaction that causes the core to melt through the containment vessel into the earth and release radioactive gases.

Sabotage or an earthquake, as well as human error or equipment failure, could cause nuclear plant crises—such as an accident in which a main pipe in the primary cooling circuit fractures. In that event, the control rods would switch off the reactor, immediately eliminating the nuclear fission process. However, the radioactive disintegration of the fission products cannot be arrested. In a nuclear power plant of 650 megawatts, the heat formation by radioactive disintegration amounts to about 200 Mw three seconds after the reactor is switched off, to 30 Mw after one hour, and to 12 Mw after 24 hours. In fact, the radioactive disintegration continues at a substantial level for several months.

Under normal operating conditions for the reactor, the external surface of the fuel casing has a temperature of about 350°C (660°F), while the interior of the fuel rods is very much hotterm as a rule 2200°C (4000°F). This temperature is approaching the melting points of the material. If the cooling liquid is lost, the outer surface of the rods would begin to heat up rapidly, owing both to the high temperatures inside and to continuous heating from the fission products. Within 10 to 15 seconds the fuel casing would begin to break down; and within a minute the casing would melt and the fuel rods themselves would begin to melt. Unless the emergency cooling of the core comes into operation in these first few minutes, the whole of the reactor core, the fuel (about 100 tons of it) and the supporting structure would all begin to melt away and to collapse on the floor of the inner most fuel tank.

To meet an accident such as this (loss of the cooling agent), the nuclear power plant is equipped with several emergency cooling systems quite independent of one another. These are based on purely theoretical calculations. The first experimental test made on them in the United States proved a complete failure; the emergency coolant could not cool the reactor core down sufficiently because it was able to escape through the leak in the cooling circuit, and owing to a layer of steam forming between the hot surface of the fuel rods and the emergency coolant, the rest of the emergency coolant was not able to carry away the heat imparted by the fuel rods. In 1978, a successful-loss-of coolant test was conducted at the Department of Energy's experimental facility. In that test loss of pressure in the coolant system triggered the reactor to shut down and the emergency system to flood the reactor core. The temperature rose to only 488°C (910°F)—meltdown would begin at 1650°C (3000°F)—and the test was over in two minutes. However, just a few months later at Three Mile Island, the emergency systems did not operate properly.

If upon the failure of the cooling system the reactor core melts, emergency coolant added at this point would only make the situation worse. The melted metals in the fuel would react violently with the emergency cooling in the circle.

# 5

# Green House Effect

Carbon dioxide ($CO_2$), a gas produced by the combustion of carboniferous substances such as coal, petroleum products and wood in industry, automobiles and households, is not directly toxic for human beings and animals, as it also if formed in every living creature by respiration and is emitted. It is harmful only in very high concentrations (asphyxiation).

Part of the carbon dioxide generated by combustion is assimilated by plants while they are bathed in light, and oxygen is given off in its place (this ratio is estimated at about 50%), or it is dissolved physically in waters. The rest of the carbon dioxide is dispersed in the atmosphere.

The oxygen-based atmosphere of the earth probably began forming two to three billion years ago by the assimilation of carbon dioxide by plant life. The carbon remaining in the plants as a result of the assimilation was deposited upon fossilization in the earth's crust and remained there in the form of petroleum, coal, natural gas, and peat.

For many decades now, the human race has been exploiting these carbon deposits to an ever increasing extent as a source of energy, and has been thus combining the carbon with oxygen again. This results in the production of carbon dioxide and the consumption of oxygen. Since industrialization began, the carbon dioxide content of the atmosphere has increased in consequence by 15%. By the year 2000 it is estimated that the increase will amount to 30%. In industrial regions the carbon dioxide contained in the atmosphere is particularly high, because not only is more of it produced there but the stock of plant life, which absorbs carbon dioxide and gives off oxygen, declines steadily.

By reason of its structure, the carbon dioxide molecule has the property of absorbing infrared rays and converting them into heat. Consequently, a slight increase in the carbon dioxide content of the earth's atmosphere considerably increases the heat-retaining capacity of the air. Normally, the earth's surface disseminates into the universe again a large part of the entering sunlight, in the form of infrared heat rays. This prevents the temperature on the earth from constantly rising under the influence of solar radiation. But if the carbon dioxide content of the atmosphere increases, a larger part of this reflected infrared radiation will be absorbed into the atmosphere, and so will remain on the earth and thereby raise the temperature.

An increase in the temperature can change the climate of the earth considerably in the long run. A rise in average temperature throughout the world by 2°—3°C (35.6°—374°F) would very probably result in the melting of large quantities of polar ice, and so lead to a rise in the sea level and to the loss of large tracts of land. That this development has not yet occurred, despite the rising carbon dioxide content of the atmosphere, is due, surprisingly, to the pollution of the atmosphere by particles of dust. Alongside the growth in the carbon dioxide content the dust content has risen simultaneously in the higher layers of the atmosphere, and solar radiation has been reduced considerably by dispersion.

The first assessments of the carbon dioxide content of the atmospehre at the beginning of the Industrial Revolution show a concentration of 280 parts per million (0.028%). By today the level has risen to 330 ppm, showing an annual growth rate of 0.07 ppm. By the year 2000 the figure is likely to rise to 379 ppm. According to present information, the overall average temperature rises by 0.5°C (0.9°F) for each increase in the carbon dioxide content by 18%. By the year 2000 an increase is consequently to be expected of 1°C (33 8°F).

**Green House Effect**

*Introduction.* The most discussed of the long term effects has been the changing concentration of carbon dioxide. This has been carefully measured at Mauna Loa in the Howaian Islands for more than a decade. Seasonal variations have been as well as a gradual increase on the average value of some 0.7 ppm each year, to reach a present value of about 320 ppm. The seasonal variation can be explained by the greater demand for carbon dioxide by plants during the spring and summer when they are growing rapidly, decreasing the amount in the atmosphere. Records of carbon dioxide concentrations over the last century have revealed that there has been occurring a gradual increase from about 290 ppm before 1890 to the present value.

Measurements at Mauna Loa, Hawaii, an area away from heavy industrial concentrations, indicate a yearly increase of approximately 0.7 ppm (during 1958-65) and 0.5 ppm from 1965. An average increase of 0.6 ppm $Yr^{-1}$ (0.2% $Yr^{-1}$) approximates to the amount of $CO_2$ formed from the burning of fossil fuels. Regular seasonal variations in the concentration of $CO_2$ correlates with photosynthetic activity, and in the

Northern Hemisphere, the concentration peaks around April, and is lowest around September-October. It is believed that the major contribution to the seasonal variation is the photosynthetic activity of mid latitude forests. Forest destruction could have a serious affect on the levels of atmospheric $CO_2$. Carbon dioxide undergoes phtochemical reactions, producing CO at higher altitudes in the atmosphere.

$$CO_2 + hv \rightarrow CO + O \quad \text{...(1)}$$

(uv)

This is a significant source of CO at these levels.

The world's climate is dependent on the global heat balance, a mean temperature change of 2-3K would have a marked effect. Of the $8.4x13^4$ $Jm^{-2}$ $min^{-1}$ (1400 watt $m^{-2}$) energy that enters the earth's atmosphere, approximately 47% reaches the earth's surface and adjacent atmosphere, directly or after being scattered. The incoming ultraviolet, visible and infrared energy has a max intensity, at the earth's surface around 483 nm, while energy re-emitted from the earth is only in the infrared region (2000 40,000 nm) with a maximum intensity around 10,000 nm. If this energy was lost to the atmosphere the temperature of the earth's surface would be around —20° to —40°C. However, some of the infrared is absorbed by water vapour and carbon dioxide in the air and re-emitted in all directions, some of which returns to the earth producing an average temperature of around 14°C.

Water has three fundamental vibration modes, $_{v1,}$ $_{v2}$ and $_{v3}$ (table 1), while $CO_2$ has four modes; $_{v1,\ v2a,\ v2b}$ and $v_3$ (modes $_{v2a}$ and $_{v2b}$ are degenerate). The strong infrared absorptions due to water occur below 8000 nm (and >20,000 nm) while the $v_2$ absorption of $CO_2$ is the most intense and absorbs over the region 13,000—18,000 nm. This leaves in infrared exit window between 8000—13,000 nm.

The earth's mean temperature rose 0.4°C from 1880 to 1940, and from 1945 has dropped around 0.1°C. The initial rise has been correlated with an atmospheric concentration of $CO_2$.

A large amount of carbon dioxide gets introduced into the atmosphere from fossil fuel burning, furnaces and breathing of animals. From fossil fuel alone, more than $2.5 \times 10^{13}$ tonnes of $CO_2$ is being emitted into the atmosphere each year. Not all of the $CO_2$ injected into the atmosphere remains there; about half of it gets utilized by plant life or absorbed by water of the oceans. Part of the $CO_2$ dissolved in the ocean may get precipitated or incorporated in marine organisms. In this respect aquatic plants in the oceans are playing an important role in maintaining $CO_2$ equilibrium between the atmosphere and the surface layers of the ocean (up to 100 metres deep). Part of the $CO_2$ taken up by terrestial plants gets deposited in dead vegetation and humus on the forest floor. Some of it, in the form of organic plant parts, has been eaten by herbivorous animals and gets deposited on or in the soil. However, much of $CO_2$ is still left in the atmosphere. An increase in atmospheric carbon dioxide will influence the photosynthesis, and consequently on plant growth by its direct fertilising effect, especially in hot tropical environments and a longer growing season in temperate regions. This potential fertilising effect should be exploitable by using modified crop varieties and agriculture practices to compensate for the disadvantageous effects of temperature increase. Carbon dioxide emitted by volcanoes during several billion years was at least 40,000 times that still present in the air. On a global time scale, the known amounts of $CO_2$ in limestone and fossil sediments suggest that normal residenc time of $CO_2$ in the atmosphere has been probably around 100,000 years.

Carbon dioxide gets confined exclusively to troposphere. In dense concentration it can act a serious pollutant. The temperature at the surface of the earth has been maintained by the energy balance of the sun's rays strike the planet and the heat that gets radiated back into the space. Some of the sun's rays that penetrate the thick layer of $CO_2$ are able to strike the earth and get converted into heat. The heated earth is able to reradiate this absorbed energy as radiations of longer wavelengths. Much of this does not pass through $CO_2$ layer to outer space but gets absorbed by the $CO_2$ and water in the atmosphere and adds to the heat that has been already present.

Thus the earth's atmosphere heats up. This phenomenon is termed as the *green house effect.* Carbon dioxide thus acts like the glass of a green house and on a global scale, tends to warm the air in the lower levels of the atmosphere.

In the "green house effect", increase in the $CO_2$ concentration means more infrared radiation is trapped and re-emitted back to the earth, producing a build up of infrared radiation in the atmosphere and a mean global temperature rise. Estimates of the rise in temperature with increasing $CO_2$ concentration range from 0.1 to 4.9°C with a mean around 2°C for doubling the $CO_2$ concentration to 600 ppm. As there is a logarithmic relationship between surface temperature and $CO_2$ concentration it is likely that a temperature will be reached, beyond which further increases of $CO_2$ levels have little effect. A maximum increase of 2.5 K has been suggested. Since 1945 the mean global temperature has dropped suggesting that either the green house effect is not a satisfactory explanation, or some other effect is beginning to dominate. Aerosols in the size range 0.1-5 um, could reduce incoming radiation by 10% through back scattering and any increase could lead to cooling of the atmosphere. The rate of increase of atmospheric turbidity has been greater than the increase on $CO_2$ levels, and it could be that this is the more important factor today. It is not yet possible to adequately explain the change in the temperature over the last centrury, but at least an increase in the $CO_2$ and aerosol content of the atmospherc has the potential to alter the world's climate. Pollutants such as $N_2O$, $CH_4$ and $CCl_3F$, which have strong infrared absorptions, could also influence the mean global temperature.

An increased heating of earth would cause recede of glaciers, disappearance of ice caps such as those found over Antarctic and Greenland, and rise in ocean level. Infact, it has been estimated that if all the ice on the earth should melt, 200 feet of water would be added to the surface of all oceans, and lowlying costal cities such as Bangkok and Venice would get inundated. Only a rise in sea level of 50 to 100 cm caused by ocean warming would be able to flood lowlying lands, in Bangladesh and West Bengal. Further, due to the much warmer tropical oceans because of increased carbon dioxide, there have been likely to be more hurricanes and cyclones and early show melts in mountains causing more floods during mansoons. According to Mr. Stephan Keckes, at Yugoslavian marine biologist, heat of the United Nations Environment Programme's Centre on ocean and coastal areas, within 30 years, rising seas will be able to wash away entire countries and flood cities from Boston to Bombay. He calculated that seas will rise by 1.5 to 3.5 metres within three decades. In Bangladesh alone 15 million people will have to move or drown. He says that little can be done about rising, oceans beyond studying the phenomenon and preparing for the worst.

According to G.N. Plass, if the carbon dioxide content of the atmosphere gets doubled, the average surface temperature of the earth would rise 6.5°F. However, this would not happen because of reflection of some of the sun's heat back into the space by densely accumulating particulate contaminants such as smoke and dust from industrial and automobile exhaust. If carbon dioxide continues to accumulate, it may disallow the cooling effect of particulate contaminants, and as a consequence the earth's temperature may rise again.

There is sufficient evidence that the temperature of the entire earth has risen slighty during recent decades. Glaciers in both hemisphere are receding. Between 1885 to 1940. there occurred an increase in mean annual temperature of about 0.9°F of the earth. After 1940, global warming subsided. Warming of Northern Europe and North America continued between 1940 and 1960, but the mean annual air temperature on global scale as well as for the Northern Hemisphere got decreased slightly (0.2°F). More than 40 per cent of the total increase in carbon dioxide content of the atmosphere from combustion occurred during that period. It might be possible that increased particulate pollutants from industry and automobile exhaust nullifed or masked any effect of temperature from carbon dioxide content of the atmosphere, but this has been not. There was an overall temperature rise of 0.7°C over the past 134 years with the three warmest years in the whole record being 1980, 1981 and 1982. In addition, five of the nine warmest years even have occurred since 1978. There have been however, some unexplained fluctuations within this long term trend, between 1980 and the mid-1980 but overall conditions were relatively steady, with land-based readings actually showing a slight fall. The latest predictions for global warming given at European meetings of climatologists and policy makers say that the temperature of the earth could increase by 1.5 to 4.5°C by the year 2050.

The trend in temperature which has taken place over the last century cannot be associated with the postulated increase in $CO_2$ concentration over the same period. Average northern hemisphere temperatures increased by 0.6°C between 1890 and 1940 but since that date have decreased by 0°C. This does not rule out the relationship between $CO_2$ concentrations and temperature but reveals the uncertainty associated with prediction of climatic change and the magnitude of natural effects of unknown cause.

It has been assumed that particulate concentrations have been increasing because of man's activities and that this increase would have an effect of the climate. Particles, for instance, increase the albedo (percentage of sunlight reflected) by scattering the incoming radiation, and cooling would be predicted as a consequence. Particles also scatter the outgoing radiation, although not so effectively because the particulate sizes that predominate. In the atmosphere are most efficient at scattering light of the incoming wavelengths.

Particles also absorb and re-emit radiation; the wavelengths at which this take palce have been a function of the particle composition. In parallel with the green house argument, if this absorption predominated in the infrared region corresponding to the earth's emission, the result would be warming and cooling would ensure on absorption of the sun's emission. Particles., unlike $CO_2$, are having a relatively short lifetime in the atmosphere, of the order of 3 to 5 days. In addition, the man-made contribution to the atmosphere's total particulate burden have been relatively small (approximately 10%). It is argued that under these circumstances it is unlikely that there may occur any change in particulate concentrations because of man's activities, and that there could not be any ssociated climatic effect. This is not to say that fluctuations in particulate loading as, for example, take place in the stratosphere following volcanic activity, do not have an influence on climate.

No melting of the ice caps has been observed yet, but some predictions are that these effects could show up in the next 30-80 years. Some melting might take place, but total conversion of the earth's ice into liquid water seems highly unlikely when the actual energy requirements are considered. For example, the energy given up when 300 litres of air is cooled 1°C (1.8°F would melt only one gram of ice if the ice was already at the melting temperature.

Some scientists feel that rather than a flood, an ice age is more likely. Their calculations indicate that the cooling of the earth's atmosphere caused by increased particulate pollution makes an ice age a possibility for the future.

Obviously, a number of variables are involved in all of these predictions, and the net effect of atmospheric pollutants on the climate and weather remain to be seen. They could prove to be very profound, or they might even cancel one another out..

### Green-House Effect in an Automobile (like car)

The internal heating due to green-house effect can be observed in a car parked with all its windows closed. This can be explained as follows. The glass windows of a car allow the visible sunlight and the very short wavelngth infra-red rays contained in sunlight to pass through them freely and go inside the car. These rays are reflected from the inside surface of the car (like dash-board, seats, etc.). Now, the infra-red radiation emitted by very hot sun and which entered the closed car was of very short wavelength, but; the infra-red radiation reflected (or emitted) by the less hot inside surface' of car is of longer wavelength. The glass windows of car do not allow this long wavelength infra-red radiation to go out though them. So, the infra-red rays get trapped inside the car. Since the infra-red radiations produce a heating effect, therefore, the interior (inside) of the car gets heated considerably. Thus, glass is a solid substance which produces green-house effect.

### Green-House Gases

Those gases which can trap infra-red radiation given by the sun to produce green-house effect leading to heating up of the environment, are called green-house gases. One of the most important green-house gas is carbon dioxide. Water vapour and ozone also have the ability to trap the infra-red radiation, so they are also called green-house gases. Thus, we have three green-house gases: (i) Carbon dioxide, $CO_2$ (ii) Water vapour, $H_2O$ and (ii) Ozone, $O_3$, Out of these three, water vapour and ozone do not contribute much green-house effect to the earth's atmosphere because ozone is present only in the upper part of atmosphere whereas water-vapour is found only near the surface of earth (which is at the bottom of the atmosphere). Only carbon dioxide contributes largely to the green-house effect in the earth's atmosphere, because carbon dioxide is much more uniformly distributed in atmosphere.

**Importance of Green-House Effect in Nature**

The green-house effect produced by carbon dioxide gas is very crucial to out existence on earth. This can be explained as follows: By producing the green-house effect, carbon dioxide gas in the atmosphere traps the infra-red rays (heat rays), leading to the heating of earth and its atmosphere. This heating of earth (or rise in temperature of earth) is very necessary for our existence because without it, the whole earth would be converted into an extremely cold planet, making the existence of life difficult.

# 6

# Air Pollution Monitoring

The Air (prevention and control of pollution) act, 1981 has in its preamble, the objective to take "appropriate steps for the preservation of the natural resources of the earth, which among other things, include the preservation of quality of air." The preservation of quality of air means the fixing up of certain minimum standards of ambient air quality in respect of the common and uncommon ingredients of air and the prevention of increase of the concentration of such ingredients in comparison to the fixed minimum quality.

These are many ingredients which may be termed as "primary pollutants." These are five primary pollutants which together contribute more than 90% of global air pollution. These are:

A. Carbon monoxide, CO

B. Nitrogen oxides, $NO_x$

C. Hydrocarbons, HC

D. Sulphur oxides, $SO_x$

E. Particulates,

After being released directly into the atmosphere, these primary pollutants react with atmospheric constituents and with themselves and thus produce secondary pollutants. In many situations these secondary pollutants are much more dangerous and injurious to the environmental quality. One of the most common secondary pollutants is 'Photochemical Smog.'

The preservation of air quality requires constant and continuous monitoring of the ambient air and effective control measures to reduce the emissions from anthropogenic sources. This also requires efficient methods to forecast the future ambient air quality and implementation of such schemes which may monitor and respond quickly and effectively to control episodal and emergency emissions. Thus, we may agree that the data collection and their interpretation play very important role in the ambient air quality preservation. Here, we shall be mainly concentrating on the sampling procedure for various air pollutants. We shall also discuss difficulties encountered in collecting representative samples and sources of error. The methods used for measuring gaseous emissions from a stack or a vent depend on the nature of the compound and the purpose for making the measurement. In addition, the composition and the temperature of the carrier gas stream affect the selection of a sampling technique, analytical method and sampling plan.

**Classification of sampling methods**

The sampling methods used for the study of air pollution can be classified under three different headings.

1. Sampling of impurities of various nature (ranging from particulate matter to gases).
2. Sampling under various environmental conditions (ranging from samples taken from chimneys to samples taken in the open air).
3. Sampling methods varying according to the time

factor (ranging from intermittent to continuous sampling).

**Difficulties encountered in sampling**

1. Collecting samples of true representative character.
2. Errors arising from methods used for the collection and separation of the various components of pollution.
3. Difficulty in preventing any change in the contentration of particulate matter in suspension, as a result of sampling operations.

**Instruments for sampling waste gases and the atmospheric sampling**

Following are the sampling devices in use:

**Devices for general use**

***Meters***

They are used to determine accurately the volume of the gas collected. They are fitted with manometers and thermometers to indicate the pressure and temperature of the gas stream sampled.

***Probes***

These are tubes suitable for penetrating into the gas stream and should be constructed of materials which are non-corrosive and which can withstand special temperature conditions. Also, they should be constructed of materials which do not react with the substances to be sampled. Therefore, they should be made of stainless steel or preferably of glass or quartz. A probe should have suitable length and diameter. To ensure isokinetic sampling conditions, the opening of the probe should face the gas stream to be sampled.

***Suction devices***

Any suction device which has the required volumetric capacity can be used. Vacuum pumps driven by electric motors are very commonly used.

**Devices for sampling gases and vapours**

***Absorbers***

In this process, effluent gases are passed through absorbers (scrubbers) which contain liquid absorbents that remove one or more of the pollutants in the gas stream. The efficiency of this process depends on

(1) amount of surface contact between gas and liquid

(2) contact time

(3) concentration of absorbing medium

(4) speed of reaction between the absorbent and gas.

Absorbents are being used to remove sulphur dioxide, hydrogen sulphide, sulphur trioxide and fluorides and oxides of nitrogen.

The equipment using the principle of absorption for the removal of gaseous pollutants includes: (1) packed tower, (2) plate tower, (3) bubble cap plate tower, (4) spray tower, and (5) liquid jet scrubber absorbers. Selective chromatographic absorption of gases on small pellets may offer much higher rates that those achieved in packed towers.

A gas can be sampled by means of suitable absorption reagent. For this purpose, U-shaped absorbers are used. These absorbers are filled with a certain measured amount of reagent and fitted with a porous glass partition, so that the air or gas led into them is passed through the reagent solution in the form of fine bubbles thus ensuring intimate contact. Sampling by means of such absorbers is usually carried out at an average rate of about 100-150 litres per hour of gas stream. The absorbers may be arranged in series of two or more elements containing two or more different reagent solutions so as to absorb different pollutants successively from the same volume of gas or air sampled.

By selecting the most suitable absorbent solutions, the gaseous components listed below can be determined in concentrations as low as 0.1 ppm by volume.

1. Oxides of sulphur
2. Oxides of nitrogen
3. Ammonia
4. Hydrogen sulphide
5. Hydrochloric acid
6. Hydrofluoric acid
7. Hydrocyanic acid.

This method can also be used for the determination of ozone, hydrocarbons and organic solvents.

***Adsorbers***

Adsorption is brought about by aspiring the air or gas to be sampled through adsorption columns containing silica gel, activated charcoal or any other suitable agent. After adsorption, the different pollutants can be extracted from the column in various ways. For example, by raising the temperature.

The main difficulty in this method is in selecting a suitable adsorbing medium. This type of sampling is used especially for ozone and light hydrocarbons.

***Condensers***

Here the gas stream sampled is cooled in suitable

containers, thus bringing about the condensation of the volatile substances present. As in the case of adsorption devices, here also the condensation traps can be arranged either in series or parallels, at decreasing temperatures. By using various coolants, e.g., ice, liquid air, or liquid nitrogen—the components can be separated by fractional condensation.

This method is used in particular for the sampling of odoriferous substances.

***Collectors under reduced pressure***

For some substances like nitric acid and aldehydes having a high molecular weight, absorption in aqueous solutions is sometimes incomplete. In such cases, it is preferable to use bottles of known volume for collecting under a pressure reduced to 200 mm Hg or even less.

To do this, the absorbent solution chosen is first introduced into the bottle and the pressure is then reduced. Then the sample is admitted until the internal and external pressures are equal and the container is shaken continuously so as to ensure maximum absorption. This method is suitable, for sampling the oxides of $N_2$.

***Plastic containers***

Special polythene bags are commonly used for collecting and transporting large volumes of air. These bags have the advantage that they can be used for successive analysis of small fractions of the sample taken. Moreover, polythene is inert with respect to many substances including $SO_2$ and formaldehyde. On the other hand, plastic bags are not suitable for collecting and storing aerosol suspensions, because of the possible generation of electrostatic charges, as a result of which the aerosols tend to move towards the walls and condense on them. Plastic bags have been widely used for grab sampling and sample storage before analysis.

**Samplers for mass-spectrometric analysis**

Sampling or mass spectrometric analysis can be carried out in various ways. For example, by compressing the gas sample in a pressure flask so as to concentrate a large quantity of gas in a small volume, or by filling evacuated containers.

**Duration of sampling period**

Two types of sampling are used in studies of air pollution. Short period or 'spot' sampling and continuous sampling for the evaluation of peak and average concentrations over definite time intervals. Spot samples are collected over periods varying from less than 30 minutes to several hours for specific, well defined purposes. The choice of sampling periods depends upon the nature of the compound under study and its stability to oxidation, light or other factors such as sensitivity, accuracy and precision of the analytical method to be used for the measurement of pollutants. Short-sampling is useful for the random checking of pollution at many points. Such samples have only limited value because pollution levels fluctuate widely, depending on meteorological conditions, typographical features, and various factors associated with sources of pollution (e.g. mass rates of emission of pollutants from smoke stack, the temperature, velocity and density of stack gases, the height of smoke stacks, the distribution of sources, and the down wind distance from the sources to the points where the measurements are made). Consequently, spot sampling cannot give adequate data on the nature and the magnitude of an air pollution problem. However, by making a series of such observations one can obtain information on the maximum and average concentration of pollutants over a definite time period. Continuous sampling techniques are usually necessary in systematic studies of the nature and extent of air pollution if the data are to be value for epidemiological surveys, for evaluating the potential hazard to man, animal or vegetation, and for control programmes. Continuous sampling may be carried out by the sequential chemical absorption or filtration of measured air volumes and subsequent analysis of pollutants, by appropriate analytical techniques.

Samples should be collected under all environmental conditions that affect pollution levels, e.g. all weather conditions. This may necessitate sampling continuously round the clock. There is sufficient evidence that pollution is influenced by diurnal variations in wind speed, atmospheric stability, solar heating and cooling, cloud cover and other factors. Biological effects, rates of corrosion, and visibility are greatly influenced by the moisture content of the atmosphere and by interaction of the water vapour with condensation nuclei. Natural air cleaning processes, including precipitation materially affect the levels of pollutants in the atmosphere.

It may be pointed out, that trace element levels in the atmosphere are generally very much lower than those in or near emitting sources. Collection of a given amount of atmospheric material therefore requires longer sampling times or higher air flow rates than is the case of source material. There are several consequences of this.

1. Very sensitive analytical procedures are generally needed to determine trace elements present in collected atmospheric samples.
2. Determination of element volume concentration is much easier than that of specific concentrations,

since the former does not require measurement of the weight of the collected material.

In order to obtain a good sample, various relevant factors as enumerated above should be carefully considered. The size of the sample required (total volume of air sampled) decreases with increasing concentration of pollutant and increasing sensitivity of the analytical method. In general, a sample of about 10 $m^3$ (cubic metres) is required. The sampling rate varies from 0.003 $m^3$/min., depending on the equipment used. The optimum sampling time for a representative sample is a 3 hour period. This definitely gives a more correct picture than the 1 hour period.

**Location of sample sites**

Sampling sites must be carefully selected so as to be representatives of the areas under study. For the study of health effects, the sites must be located so that the collected samples represent air that is actually breathed by the exposed population groups. Since the concentration of pollutants varies with altitude, results obtained on the roofs of tall buildings may differ substantially from conditions at the ground or breathing level. The necessary number of sampling stations, and their location, depend on several factors including the objectives of the programme, the size of the study area, the proximity of the sources of pollution, typographical features and the weather. A representative number of sampling stations for a given area may be established by means of a preliminary survey, whose objectives should be:

1. To gather information on the nature and magnitude of the emission from principal sources of pollution.

2. To review the available climatological and meteorological data.
3. To gather data on the concentration of pollutants in areas of severe and slight pollution.

**Sampling of particulates**

Filtration is the most common technique for sampling particulate matter. The filter materials are (a) cellulose filter paper for determination of metals, (b) glass-fibre for analysis of organic compounds, and (c) silica felts for trace inorganic species and organic compounds. The membrane filter is quite common—it yields high flow rates with small, moderately uniform pores. The optimum pore size of these filters is 0.45m.

(i) High volume air samplers are used to pump large volumes of air, up to 2000 $m^3$, at a rate of about 1.7 $m^3$/min. The filters used in a high-volume sampler usually consist of glass fibres and have a collection efficiency of more than 99% for particles with 0.8m diameter. Particles with diameters exceeding 100m remain on the filter surface, whereas particles with diameters less than this to about 0.1m are collected on the glass fibres in the filters. For efficient collection of particles, one should use very small diameter fibres (less than 1m) for the filter material.

From the environmental health viewpoint, it is desirable to draw air samples where people are working. For this purpose, lightweight battery-operated pumps are used, which can pump air at 2 lit. $min^{-1}$ through a 3.7 cm glass-fibre paper for upto 8 hours. Samples of polluted air are collected at several points to build up a picture of the distribution of an air pollutant. Samples collected by filters are analyzed chemically, by weighing, by microscopes and by particle sizing.

(ii) *Sedimentation* is a simple technique for collection of particles having diameters exceeding of 5m. The apparatus may be a simple glass jar fitted with a funnel. A liquid is often added to the collector so that solids are prevented from being blown out by air.

(iii) *Electrostatic* samplers are very efficient for collection of small particles. These are also smaller versions of pollutant control devices. On entering the sampler, particles pick up charge as they pass through an electrical discharge between two electrodes maintained at a potential difference of up to 30,000 volts. The small charged particles in a gas stream lose their charge in contact with an electrode (oppositely charged) and accumulate on the electrode.

(iv) *Thermal precipitators* may be used for collection of aerosol particles of 0.001m size, with high efficiency. They work on the principle that suspended particles move to lower temperature regions when exposed to a high temperature gradient.

(v) *Impingers* collect particles from a relatively high-velocity air stream directed at a surface. The device is a dry impinger when the collecting surface is dry, and wet impinger when the collecting surface is wet. Dry impingers of the Cascade impactor type can differentiate particles according to their size. The cascade impactor directs the air stream against collection slides through increasingly smaller orifices and provides for stepwise collection of smaller particles. However, sometimes wrong results on the high side are likely to be obtained for small particles levels since large particles may break up into smaller pieces from the impact of impingement.

**Sampling suspended particulates by high volume filtration (The High Volume Sampler)**

The high volume sampler is one of the common instruments used for the collection of suspended particulate matter in recent years. Stoke's law offers the basic approach to collection by this technique.

When operated in virtually quiescent air, the geometry of the high volume sampler employs the sloping roof of the shelter as a means for causing air entering the sampler under the caves of the roof to change direction by at least 90° before entering the horizontal filter. Particles that remained entrained in the air sample prior to horizontal filtration have, in so doing, satisfied the definition of truly suspended dust, i.e. dust that is not subject to settling under the influence of gravitational force.

**Air sampler operation**

The high volume sampler is a vacuum cleaner type motor that is used to draw a sample through a filter area. The filter most commonly used is the 20 cm x 25 cm mat, which allows collection of an air sample at a rate from 40-60 cfm (18,900-28,350 cc/s) over a nominal sampling period of 4-6h, and a normal sampling period of 24 h. These conditions permit the sampling from 45000-75000 c ft. (1260-2100 $m^3$) of ambient air, with consequent extraction of about 0.5 g of suspended particulate (aerosol). This provides quite a substantial weight of sample, which greatly simplifies subsequent chemical or physical analysis.

The sampler unit may be divided into three parts. They are:

(1) the face plate and gasket,

(2) the filter adapter assembly, and

(3) the motor or blower unit.

It is installed inside a casing that is provided with a roof so that the filter is protected from precipitation.

The motor, capable of continuous operation for 24 hours with input voltage ranging from 20-230 V, is usually started and stopped by a simple clock timer. The duration of sampling is measured in an elapsed time meter that is placed in series with the blower. Starting and finishing time are at the discretion of the operator, although some recommend starting and finishing from midnight to midnight-24 hours over the same calendar day. The national Air Sampling Network, in the United States, operates such samplers over the entire country on scheduled 24 h intervals every sixth day. This schedule permits a sampling of everyday of a week over a period of seven weeks and is the recommended approach for long-term air sampling in a community. On the other hand, short-term studies to determine day-to-day variations in particulate levels may require continuous daily 24-hour sampling. Fulfilment of the aims of such as programme, without manual changing of filters at midnight, is possible by:

1. Using one sampler and manually changing the filter at noon or at some other convenient hour rather than at midnight.

2. Using two high volume samplers side by side with a timber switch set to start the second motor as the first stops at midnight. The exposed filter can then be changed conveniently during the daylight hours while the second 24 hour sample is being collected.

An airflow measuring device, such as a rotameter, is used to record the volume of sample air passing through the filter for later use in calculating the mass of particulate per static volume of air (micrograms per cubic metre of sample air). The initial airflow, through the clean filter mat is first measured. At the completion of a 24-h (or other sampling time), another rotameter reading is made on the exposed filter, and the average flow is used in the calculations.

**Filters used in high volume air sampling**

The use of common laboratory analytical grade paper filters is not recommended for sampling of ambient air as they are not satisfactory for the recovery of very small particles (< 0.5m). In addition, such paper filters often do not possess the mechanical strength to withstand airflows of the order of 25 lit/s.

Several types of filter materials have been used and are currently commercially available for collecting atmosphere particulates. Probably the simplest and cheapest is a dense cellulose paper such as what man No. 41 analytical filter paper. This has sufficient mechanical stability to enable the use of the paper for high volume sampling. Paper does, however, have a broad distribution of pore sizes and can vary considerably in density, thus collection inefficiencies are observed for particles sizes below 0.3m when sampling is conducted for short periods. This problem is somewhat overcome by pore clogging effects when 24 hour sampling periods are employed. A significant disadvantage in the use of filter paper is its tendency to absorb water. This necessitates careful equilibration of the paper to standard relative humidity, if accurate weights of collected material are required.

Several forms of organic membrane filters have come into use for particle collection during the last few years. The most common are Millipore, Nucleopre, Pallflex and Acrapor, which are composed of cellulose acetate or nitrocellulose, and Fluoropor and Miller, which are Teflon filters. In general, organic membrane filters having high flow resistance must be used in conjunction with a relatively pressure insensitive pump. Cellulose membrane filters are mechanically fragile and are best suited to analyses that can be directly performed on the filter. They are readily soluble in most common organic solvents for ease of removal of collected particles. Millipore filters are slightly hygroscopic and should be equilibrated at constant humidity for precise weight determination. Inorganic impurity levels in Millipore filters are generally low. Pallflex and Acrapor exhibit impurity levels that are substantially high for the determination of many collected particulate elements. Teflon filters, are inert to both acids and organic solvents and therefore would permit multiple use.

Glass fibre filters are the most widely used materials for the collection of airborne particles. Despite mechanical properties, which are inferior to those of cellulose paper, they are suitable for use in high volume samplers.

Furthermore, a glass fibre filter is inert to acids (excepting hydrofluoric acid) and to organic solvents. For accurate determination of collected weight, filters must be equilibrated in an atmosphere at constant relative humidity. Unfortunately, glass fibre filters contain substantial amounts of several trace elemental impurities (viz. Cr, Cu, Fe, Mn, Ni, Pb, Sb and Zn). Consequently, in order to reduce the contribution of impurities by the filter it becomes necessary to collect large amount of sample.

#### Stack sampling techniques

Stack sampling or source sampling may be defined as a method of collecting representative samples of pollutant laden air/gases at the place of origin of pollutants to determine the total amount of pollutants emitted into the atmosphere from a given source in a given time. Stack sampling is employed for the assessment of the following:

1. To determine the quantity and quality of the pollutant emitted by the source.
2. To measure the efficiency of the control equipment by conducting a survey before and after installation.
3. To determine the effect on the emission due to changes in raw materials and processes.
4. To compare the efficiency of different control equipments for a given condition.
5. To acquire data from an innocuous individual source so as to determine the cumulative effect of many such sources.
6. To compare with the emission standards in order to assess the need for local control.

Source sampling is usually carried out in a process ventilation stack to determine the emission rates and/ or characteristics of pollutants.

**Planning the study**

The success of stack sampling depends on proper initial planning prior to the conduction of survey. The planning includes the following:

1. Familiarity of the process and operations to determine the time of cyclic operations, peak loading that might cause variations in the characteristics.
2. Methods of sampling
3. Methods of analysis of samples
4. Sampling time because certain industries undergo cyclic changes.

5. Amount of samples required
6. Sampling frequency

**Representative sample**

It is highly important that the sample collected must truly represent the conditions prevailing inside the stack. The important considerations for accurate representative sample collection include:

1. Accurate measurement of pressure, moisture, humidity and gas composition.
2. The selection of suitable locations for sampling.
3. Determination of the traverse points required for a velocity and temperature profile across the cross-section of the stack and sampling particulate matter.
4. The measurement of the rate of flow of gas or air through the stack.
5. Selection of a suitable sampling train.
6. Accurate isokinetic sampling rate especially for particulate sampling.
7. Accurate measurement of weight and volume of samples collected.

**Selection of sampling location**

The selection of the best sampling point is almost invariably a matter of compromise and individual judgment.

The sampling point should be as far as possible from any disturbing influence, such as elbows, bends, transition pieces, baffles or other obstructions. The sampling point, wherever possible should be at a distance 5-10 diameters down stream from any obstructions and 3-5 diameters up-stream from similar disturbance.

**Size of sampling point**

Usually, there will not be any opening in the stack. Hence for collection of samples, an opening has to be made to an extent of accommodating the probes. The size of sampling point may be made in the range of 7-10 cm, in diameter. A flange may be riveted so that the opening may be closed during the non-sampling period.

**Traverse points**

For the sample to become representative, it should be collected at various points across the stack. This is essential as there will be changes in velocity and temperature (hence the pollutant concentration) across the cross-section of the stack. Traverse points have to be located to achieve this. These points are to be located at the centre of each of a number of equal areas in the selected cross-section of the stack. The number of traverse points may be selected with reference to Table 1.

**Table 1**
*Traverse Points*

| Cross-section area of Points Stack sq. m | No. of |
|---|---|
| 0.2 | 4 |
| 0.2 to 2.5 | 12 |
| 2.5 and above | 20 |

In circular stacks, traverse points are located at the centre of equal annular areas across two perpendicular diameters. Table 2 may also be referred to for fixing the traverse points and the distances are from the inside wall of the stack.

In case of rectangular stacks, the area may be divided into 12 to 25 equal areas and the centres for each area are fixed. The traverse should be carried out at least on nine hypothetical squares on at least three lines.

**Table 2**

*Location of Traverse Points in Circular Stacks*

| Traverse point number on a diameter | Number of traverse points on diameter | | | | | | | | | | |
|---|---|---|---|---|---|---|---|---|---|---|---|
| | | 6 | 8 | 10 | 12 | 14 | 16 | 18 | 20 | 22 | 24 |
| 1 4.4 | 3.3 | 2.5 | 2.1 | 1.8 | 1.6 | 1.4 | 1.3 | 1.2 | 1.1 | | |
| 2 | 14.7 | 10.5 | 8.2 | 6.7 | 5.7 | 4.9 | 4.4 | 3.9 | 3.5 | 3.2 | |
| 3 | 29.5 | 19.4 | 14.6 | 11.8 | 9.9 | 8.5 | 7.5 | 6.7 | 6.0 | 5.5 | |
| 4 | 70.5 | 32.3 | 22.6 | 17.7 | 14.6 | 12.5 | 10.9 | 9.7 | 8.7 | 7.9 | |
| 5 | 85.3 | 67.7 | 34.2 | 25.0 | 20.1 | 16.9 | 14.6 | 12.9 | 11.6 | 10.5 | |
| 6 | 95.6 | 80.6 | 65.8 | 35.5 | 26.9 | 22.0 | 18.8 | 16.5 | 14.6 | 13.2 | |
| 7 | | 89.5 | 77.4 | 64.5 | 36.6 | 28.3 | 23.6 | 20.4 | 18.0 | 16.1 | |
| 8 | | 96.7 | 85.4 | 75.0 | 63.4 | 37.5 | 29.6 | 25.0 | 21.8 | 19.4 | |
| 9 | | | 91.8 | 82.3 | 73.1 | 62.5 | 38.2 | 30.6 | 26.1 | 23.0 | |
| 10 | | | 97.5 | 88.2 | 79.9 | 71.7 | 61.8 | 38.8 | 31.5 | 27.2 | |
| 11 | | | | 93.3 | 85.4 | 78.0 | 70.4 | 61.2 | 39.3 | 32.3 | |
| 12 | | | | 97.9 | 90.1 | 83.1 | 76.4 | 69.4 | 60.7 | 39.8 | |
| 13 | | | | | 94.3 | 87.5 | 81.2 | 75.0 | 68.5 | 60.2 | |
| 14 | | | | | 98.2 | 91.5 | 85.4 | 79.6 | 73.9 | 67.7 | |
| 15 | | | | | | 95.1 | 89.1 | 83.5 | 78.2 | 72.8 | |
| 16 | | | | | | 98.4 | 92.5 | 87.1 | 82.0 | 77.0 | |
| 17 | | | | | | | 95.6 | 90.3 | 85.4 | 80.6 | |
| 18 | | | | | | | 98.6 | 93.3 | 88.4 | 83.9 | |
| 19 | | | | | | | | 96.1 | 91.3 | 86.8 | |
| 20 | | | | | | | | 98.7 | 94.0 | 89.5 | |
| 21 | | | | | | | | | 96.5 | 92.1 | |
| 22 | | | | | | | | | 98.9 | 94.5 | |
| 23 | | | | | | | | | | 96.8 | |
| 24 | | | | | | | | | | 98.9 | |

**Isokinetic Conditions**

The efficiency of the sampling depends on the conditions at which sampling was carried out. The sample collected must be representative like a composite waste water collection. This can be achieved by isokinetic conditions which exists when the velocity in the stack $v_8$ equals the velocity at the top of the probe nozzle $v_n$ at the sample point. This is especially important when the particle size is greater than 3m as it presents a problem in that, the inertial effect on the particles can result in erroneous samples as explained below.

As illustrated in Fig. 5 when the velocity of the gas within the sampling nozzle is less than the gas velocity in the duct, portions of the gas stream approaching at a higher velocity are deflected. This results in the deflection of the light particles to follow the deflected gas stream and they do not enter the problem. The heavier particles by virtue of their inertia, continue into the probe with the result that a non-representative high concentration of coarse particles is collection and the sample weight is in error in higher side. Conversely, when the velocity in the probe is higher than that of the gas stream being sampled, a convergent air stream will develop at the nozzle face, with an excessive amount of lighter particles entering the probe. This results in high concentration of lighter particles and hence the sample weight is on the lower side.

**Determination of Gas Composition**

The first step in the field work of stack sampling is to determine the gas composition. This can be determined by using Orsat apparatus. The gas is collected in the Orsat apparatus and analysed for the composition of $CO_{2,}$ $O_2$ and CO in the same order, and the remaining is assumed to be nitrogen.

Molecular weight of gas = $m_x$ $B_x$

where, $M_x$ is the molecular weight of $CO_2$, $O_2$, CO and $N_2$ (44,32,28 and 28 respectively) and $B_x$ represents % of gases.

**Determination of Moisture Content**

The moisture content in the stack may be determined by any one of the following methods:

1. Wet bulb and dry bulb temperature technique
2. Condenser technique
3. Silica gel tube

The wet and dry bulb technique is used when the moisture content is less than 18% and dew point is less that 51°C and cannot be used for acid streams. As the moisture content in the stack is usually less than 18%, the wet bulb and dry bulb technique is being adopted.

**Determination of Temperature**

The temperature has to be measured across the cross section of the stack at predetermined traverse points. The temperature probe is inserted into the stack and the readings are taken with the help of a pyrometer. Care should be taken in selecting the proble.

| Type of probe | Temperature range |
|---|---|
| Chromel/Alumel | 148.8-1260°C |
| Copper/Constantan | 148.8-348.9°C |
| Iron/Constantan | 115.5°C-1010°C |
| Platinum/Platinum 10% & Rhodium | 0°C-1537.7°C |

**Determination of Velocity**

Velocity measurement is the most essential part of stack sampling. This measurement is not only required for establishing the gas flow rate but also for fixing isokinetic conditions.

There are many measuring devices available. The standard pitot-tube in combination with a differential manometer is the most widely used.

### Gaseous Sampling

The procedure adopted for gaseous sampling in a stack differs little from that employed in particulate matter sampling in so far as basic routine is concerned. Isokinetic sampling principles are not required. However, the success of gaseous sampling depends on the proper collecting device, since different gases, mists and vapours exhibit a wide variety of physical and chemical characteristics.

### Methods for Collection of Gaseous Samples

Gaseous sampling may be carried out by any one of the following methods:

1. Absorption
2. Adsorption
3. Freeze out/condensation

Whenever the flow rate and the process operation are uniform and whenever analytical equipment of extreme sensitivity is available it is possible to take a grab sample from the stack by any one of the following methods:

1. Use of an evacuated container
2. Purging (displacement of air)
3. Inflation of flexible bag
4. Use of syringe

### Proportional sampling

Whenever the source conditions, especially flow rate changes, with time, the sampling must be done proportionally so that the sample is representative. This requires the velocity measurement as explained and correction to standard conditions.

In India, not too many sophisticated instruments are available and lot of research is needed to modify the instruments that are presently being used. Table 4 gives the list of selected sampling and monitoring instruments developed or manufactured in India. The names of the manufacturers or organizations where they were developed are not given. Only the names of the instruments, the parameters and principles of operation are given. National laboratories, national institutes, research laboratories and industries are all actively engaged in the development of more dependable and accurate pollution-measuring instruments. The day is not far when we will be in a position to measure pollution accurately and adhere to the standards laid down for liquid and gaseous effluents from process industries. One can also hope for more dependable monitoring of atmospheric air, identification of sources of pollution and taking of appropriate control measures. All this would require instruments and monitoring equipment that are sound in every way—principle engineering and reproducibility-wise. Today we should lay more stress on the reproducibility of measurements made by instruments and devices, independent of any bias due to person, place or pollutant.

# 7

# Air Pollution Control

Air pollution is a topic which, among many others, has been at the forefront of social concern for the past several years. Most of this concern, as far as the general public is concerned, has been directed towards the health, distributional and regulatory aspect of the problem. To all appearances, the public has given little thought to the technological aspects of reducing emissions into the atmosphere. The term air pollution control is capable of two interpretations. To the public at large it means probably the limitation or prohibition of emissions by force of low. Inherent in this interpretation is the determination of which substances should be limited and what extent they should be limited, which requires determination of the effects of each substance on health, damage to property, and aesthetic values. The interaction of different pollutant substances must also be considered. These areas of air pollution control have been extensively explored in the past several years.

We shall be dealing with the various techniques of air pollution control, and the word control here, is used in the sense of prevention. What means are available to prevent air pollution from occurring? Aside from shutting down all polluters, which would mean disabling our economy and disrupting our way of life there are means available or potentially available to remove all pollutants to the extent necessary to prevent serious atmospheric contamination. These means take the form of certain devices which are subject to engineering analysis, and which can be designed, built, installed, and operated to achieve the desired result, but not without problems. Air pollution control is a very challenging and enterprising job considering the diversity of problems faced in regard to the number of pollutants and their quantities in the emission gas, the point at which these emissions are taking place (inadvertently and/or advertently) and the variation of the quality of emission for various process industries and plants.

The adoption of a particular control technique depends to a large extent on the quantity of the materials to be handled or the capacity of the plant, the type and properties of the materials handled, the unit operation and unit process involved in the plant, the product specification and depending on this, the purification schemes, the operating and maintenance practices in vogue, the availability of space at the plant site (in old plants) and the cost benefit analysis of the system. However, one single factor which has overbearing influences on the control technique to be adopted, is the air pollution control regulatory laws enforced in the region. (The role and will of the regulator agency to withstand the political pressures and the money power are of paramount importance).

**Control of gaseous contaminants**

The control of gaseous pollutants from stack gases depends on their properties. The methods of control include

1. Combustion
2. Absorption
3. Adsorption

4. Closed collection and recovery systems
5. Masking and counter action (for odours)

**Combustion**

Combustion processes like flame combustion or catalytic combustion can be utilized to greatest advantage when the gases or vapours to be controlled are organic in nature. Equipment employing the principle of flame combustion include

(a) Fume and vapour incinerators

(b) After-burners

(c) Flares, either with steam injection or venturi flare

The use of after-burners on incinerators has met with varying success depending on the kind of after burner used and the type of incinerator. Flare design should provide for smokeless combustion of gases of variable composition and a wide range of flow rates. Venturi flares mix air with the gas in the proper ratio prior to ignition to achieve smokeless burning. Steam injection flares mix steam with the stack gases as they reach the stack.

When the concentration of the combustible portion of gas stream is below flammable range and when lower operating temperatures are desired, catalytic combustion process are used. Catalytic combustion process is used with success for the control of effluent gases, fumes and odours from refineries, burning waste, cracking gases, phenolicres in curing ovens, paint and enamel ovens, coffee roasting processes, foundry core baking ovens and chemical plants discharging maleic and pathalic anhydrides. Gases and fumes containing excessive amounts of particulate matter reduce the effectiveness of catalytic combustion units due to a coating that forms on the catalyst.

***Major limitations***

The method is too expensive when,

1. Fuel values of the gaseous discharge are low
2. Moisture content of the discharge is high
3. Exhaust volume is extremely large

However, combustion control equipment has been used with an advantage, in petrochemical, paint and varnish and fertilizes industries.

Factors to be considered in the design of the incinerator are

1. Sufficient air ($O_2$) for combustion reaction
2. Adequate temperature for continuous oxidation
3. Adequate retention time

Generally, incineration equipment consists of a single combustion chamber with provisions for maintaining continuous and intimate mixing of the contaminant-laden gas and the flame from the gas burner. The combustion chamber is so proportioned that the gas velocity and gas flow patterns established will produce adequate retention time in the combustion zone. Average retention time is 0.2-0.3 sec. at temperatures of 650°C and higher.

**Absorption**

In this process, effluent gases are passed through absorbers (scrubbers) which contain liquid absorbents that remove one or more of the pollutants in the gas stream. The efficiency of this process depends on

1. Amount of surface contact between gas and liquid
2. Contact time
3. Concentration of the absorbing medium

4. Speed of reaction between the absorbent and the gases.

Absorbents are being used to remove $SO_2$, $H_2S$, $SO_3$, F and oxides of Nitrogen. The absorbents may be either reactive or non-reactive with the pollutants removed by them. Some of the reactive absorbents are regenerative (i.e. they may be treated and reused), while others are of non-regenerative type.

The equipment using the principle of absorption for the removal of gaseous pollutants include

(a) Packed tower

(b) Plate tower

(c) Bubble-cap plate tower

(d) Spray tower

(e) Liquid jet scrubber absorbers

Selective chromatographic absorption of gases on small pellets may offer much higher rates of absorption than those achieved in packed towers.

The absorbents commonly used for different gases are given in Table 1.

**Table 1**

*The Common Absorbing Solutions Used for Removing different Gaseous Pollutants from Gas Streams*

| Gaseous pollutant | Common absorbents used in solution form |
|---|---|
| $SO_2$ | Dimethylaniline, mixture of xylidine and water (1:1), ammonium sulphite, basic aluminium sulphate, ethanol amines (mono ethanol amine, diethanol amine, methyl diethanol amine or triethanol |

| | |
|---|---|
| | amine), sodium sulphite, ammonium sulphate and bisulphite, water, alkaline water, a suspension of $Ca(OH)_2$, calcium sulphite and calcium sulphate, barium thionates and sulphites. |
| $H_2S$ | NaOH and phenol mix (Mole ratio 3:2), tripotassium phosphate, sodium alamine or potassium dimethyl glycine, ethanolamines, soda ash solution containing suspended iron oxide or hydroxide, soda ash alone, sodium thioarsenate, ammoniacal liquor from coke ovens. |
| HF | Water, sodium hydroxide |
| Oxides of nitrogen | Water, aqueous nitric acid. |

***Packed tower***

A packed tower consists of a vertical shell filled with a suitable packing material and the liquid flows are being improved in recent years by use of new kinds of packing materials.

***Plate tower***

A plate tower consists of a vertical shell in which are mounted a large number of equally spaced circular perforated plates; gases and vapours bubble upward through the liquid seal above each plate.

***Bubble-Cap plate tower***

This consists of a vertical shell in which are mounted a large number of equally spaced circular bubble cap plates. In a spray tower, the absorbing liquid in sprayed through the gas. By this contact of gas and liquid is possible.

Efficiency of gas absorption depends upon the amount of surface contact between the gas and the liquid, time of contact between gas and liquid, concentration of the absorbing liquid, and the rate of reaction between the gas and the absorbent.

Sometimes, the carrier gas may contain particulate matter. In such a case the gas absorption equipment should also function as a wet dust collector.

**Adsorption**

In this process the effluent gases are passed through adsorbs which contains solids of porous structure. The commonly used adsorbers include activated carbon, silica gel, activated alumina, lithium chloride accivated bauxite, etc. Active carbon appears to be the adserbent most suitable for recovering organic solvent vapours. The steps necessary for effective removal of gaseous pollutants are

1. Contact of the gaseous or vaporous pollutant with the solid adsorbent
2. Separation (desorption) of the adsorbed gaseous pollutant from the solid adsorbent by regeneration or replacement of the adsorbent
3. Recovery of the gases for final disposal

The efficiency of removal of gases by adsorbents depends on

(a) The physical and chemical characteristics of the adsorbent

(b) The concentration and nature of gas to be adsorbed

Desorption is accomplished by raising the temperature of the granular bed above the boiling temperature of the pollutant by superheated steam, submerged heating elements or combustion gases. Desorption may also be performed by reducing the pressure. The adsorbents commonly used for removal of different gases are given in Table 2.

**Table 2**
*Common adsorbents Used for Removing Different gaseous Pollutants*

| Gaseous pollutant | Adsorbents used in solid form |
|---|---|
| $SO_2$ | Pulverized limestone or dolomite, alkalized alumina (aluminium oxide plus sodium oxide) |
| $H_2S$ | Iron oxide |
| HF | Lump limestone, porous sodium fluoride pellets |
| Oxides of nitrogen | Silica gel |
| Organic solvent vapours | Activated carbon |

Adsorption is a surface phenomenon and requires very large solid surface areas to be effective. In almost all cases these areas are internal as with porous material and may be almost unbelievably great per unit of adsorber volume (e.g. 1 cuin. of activated carbon has a surface area of 20,000 sq. yds). Adsorption is primarily a physical phenomenon.

Adsorption as a means of air pollution control, offers certain advantages for some classes of contaminant. It includes gases and vapours for which other means of collection are uneconomical, hazardous or impossible. For example, many flammable organic compounds may be recovered with relatively high efficiency by adsorption at concentrations are very low, usually toxic or odorous compounds can be sometimes be removed by adsorption on activated carbon, when all other methods fail.

Adsorption equipment is generally the packed bed of some adsorbing material. Installation costs for adsorbers are high but maintenance and operating costs are not excessive. Sometimes, the value of the recovered material enhances economic feasibility.

**Closed circuit and recovery systems**

Gases like $SO_2$, Oxides of N, and hydrocarbons can be recovered from the waste gas streams if they are present in sufficient concentrations. For example, where the concentration is of the order 5-10% $SO_2$ as in smelter gases, the sulphur content may be recovered economically. The most usual method at smelter is to use the $SO_2$ stream as the raw material for the manufacture of $H_2SO_4$. Similarly, the vapour recovery methods used in refineries are useful when the concentration of hydrocarbons in the effluent stream is high and relatively uncontaminated.

Oxides of N from waste gas streams in a $HNO_3$ plant are recovered using commercial zeolite. Oxides of N absorbed in the bed are recovered as enriched oxides of N and $HNO_3$ by regenerating the bed at elevated temperatures with hot air of stream.

In the alkalized alumina absorption process, oxides of sulphur in the stack gas are absorbed on spheres (1.6 mm) of alkalized alumina (a mixture of aluminium oxide and sodium oxide) in a bed suspended in the gas stream. The oxides are then removed from the spheres by reaction with a reducing gas containing hydrogen and CO, producing $CO_2$ and $H_2S$. The $H_2S$ is converted to elemental $S_2$ which can be sold and the regenerated alkalized alumina is recycled. The process would remove about 90% of the oxides of S in the stack gas. On a 800 MW power plant burning coal of 3% sulphur content, it would produce about 180 t of S per day.

In another process known as wet lime process for removing $SO_2$ from power plants, pulverized limestone is injected into the boiler furnace, where the heat drives off $CO_2$ converting the $CaCO_3$ to the reactive oxide form. The oxide then reacts with the $SO_2$ to form solid sulphites and sulphates. Some of the conversion takes place before the stack gas reaches the water scrubber, but most of it takes place in the scrubber after the reactions dissolve in the water. The resulting solids, as well as the fly ash removed in the scrubber, go to the settling pond, and water from the settling pond is recycled to the scrubber.

**Odours and their control**

Odour is undoubtedly the most complex of all the air pollution problems. Odour is caused in many cases by very minute quantities of substances. The only good measuring device for odour is the human nose, which as we all know is undependable. Further, people have mixed reactions with respect to the offensiveness of odours. Perfume liked by some, may be disliked by others!

Odours are emitted by many industries. Food processing, oil refining, paper and rubber industries and tanneries are among the major odour-emitting industries. Though foul smells may not cause direct damage, they are as much as a nuisance as noise, dirt or corrosion. The primary effect of disagreeable odours on people is the nuisance effect. Secondary effects, in some cases, may be nausea, insomnia and discomfort. On the economic front, the loss of property values near poorly operated dumping yards and slaughter houses is partly a consequence of offensive odours.

Until recently, very little attention was given to the control of odourous air contaminants. Consequently, the field of odour control has remained unexplored to some extent.

**Background information about odours**

In tackling the problems of odour control, it is necessary to know something about the background of odours and the reaction of the individual to odours.

Odours can be defined as the 'perception of small.' The experience of smell can be taken to mean any perception that results from nasal inspiration, or may refer to sensations received through the receptors of the olfactory epithelium, situated within the upper nasal cavity.

At present there is no instrument or means of chemical analysis that can replace the function of the nose in its ability to detect and to decide the state-rating of odours. Even otherwise, the development of such a unit or means of analysis would not necessarily solve the problem, since the perception of odours is an extremely subjective function.

Following are the well known facts about odours:

1. Some substances are odourous and some are not.
2. Substances of similar or dissimilar chemical constitution may have similar odours.
3. Nature and strength of the odour may change on dilution.
4. Weak odours are not perceived in the presence of strong odours.
5. Odours of the same strength blend to produce a combination in which one or both may be unrecognizable.
6. Constant intensity of odours causes an individual to quickly loose awareness of the sensation and only notice it when it varies in intensity.
7. Fatigue for one odour may not affect the perception of dissimilar odours but will interfere with the perception of similar odours.
8. An unfamiliar odour is more likely to cause complaint than a familiar one.

9. Two or more odourous substances may cancel the smell of each other.
10. Odours travel downwind.
11. Man can smell at a distance.
12. Many animals have a keener sense of olfaction than a man.
13. Likes and dislikes often depends on the association of the scent with pleasant or unpleasant experiences.
14. The number of different and distinct smells is great.

**Source of odours**

An odorant generally originates from a solid, a liquid, or a concentrated gas. Odour sources may be confined in space, like emission from ducts, or they may be unconfined, like drainage ditches, and settling lagoons.

Hydrogen sulphide, carbon disulphide, mercaptans, products of decomposition of proteins (especially those of animal origin), phenols, and some petroleum hydrocarbon are the malodours which are very common.

Odourous compounds are also generated due to various human activities. Garbage dumps, sewage works and agricultural activities are typical examples. Bad smell is also produced by decaying vegetation. The exhaust from motor vehicle is also a common source of malodour.

The sources of odour are so many, that it is almost impossible to prepare a complete list of them.

Table 3 indicates the various odourous industrial operations.

**Table 3**
*Odourous Industrial Operations*

| No. | Industry | Odourous Material |
|---|---|---|
| 1. | Pulp and paper | Mercaptans |
| 2. | Tanneries | Hides, flesh, hair |
| 3. | Fertilizer | Ammonia, nitrogen compounds |
| 4. | Petroleum | Sulphur, compounds from crude oil, cresols |
| 5. | Chemical | Ammonia, phenols, mercaptans, hydrogen sulphide, chlorine, organic products |
| 6. | Foundries | Quenching oils |
| 7. | Pharmaceutical | Biological extracts and wastes, fermented |
| 8. | Food | Cannery waste, dairy waste, meat products, packing house wastes, fish, cooking odours, coffee roaster effluents |
| 9. | Detergent | Animal fats |
| 10. | General | Burning rubber, solvents, incinerator smoke |

**Measurement of odour**

The measurement of odour falls into two categories.

1. Determination of the threshold concentration (minimum identification level) of odoriferous gases.
2. Determination of the type and intensity of atmospheric odours.

**Determination of the threshold concentration**

For each odour there exists a concentration below which no perception is possible for most individuals. This concentration is called the threshold concentration or minimum identification level.

Most of the odour-measuring devices (osmometers) are based on the vapour-dilution method. In this method, the amount of odour producing substance added to purified air is generally increased until an observer or 'sniffer' can just detect the odour.

The odour threshold for acetone is reported to be 100 ppm, acetic acid 1 ppm and for pyridine 0.02 ppm.

Generally, odour intensity increases with the odorant concentration. The relationship between odour intensity and concentration can be expressed as

$P = K \log S$ where $P$ = odour intensity
$K$ = constant
$S$ = odour concentration

Thus, we see that the odour intensity varies as the logarithm of the concentration.

**Determination of the type and intensity of odours**

For odour intensity measurements generally a panel of 6 to 12 persons in normal health are employed. The panel members sniff the air at a given location at the same time and report individually the nature and intensity of the odour. Some experience may be required to identify an odour. The odour intensity is usually stated according to a predetermined rating system.

A widely used scale for odour intensity is the following: O, no odour; 1. threshold level (barely perceptible) 2. Definite odour; 3. strong odour; and 4. overpowering odour. Half scores may be used when the observer is undecided. By averaging the values recording by members of the panel, a single value can be assigned to the odour intensity at a given location. If measurement are made by the same panel at various locations, the source of the particular odour may sometimes be found.

Several descriptive words are used to characterize the type or quality of an odour. For example fruity, flowery, burnt, aromatic, fragrant, sweet, garlic, foul, repulsive, nauseating, etc. Sometimes lay terms are used for odour identification. For example acetic acid has a vinegary odour, hydrogen sulphide has an odour of rotten eggs and sulphur dioxide has an irritating odour.

The expert panel method is fairly expensive and limited to short periods of use. Also, care must be taken because mixture of two or more different odours can be very misleading. Thus, the identification and determination of odours in the open air is even less precise than the determination of threshold values in the laboratory.

Another method used by control officers, is to have residents in certain areas to note the time of day when a particular odoured is notice and the number of times over a specific period. This data may be collected and reported graphically.

For determining the concentration of odoriferous compounds in ducts and stacks, a syringe technique is employed. Measured samples of odour-laden gas are taken in hypodermic syringes and brought to an odour-free room. A small quantity of this gas is then added to a syringe into which odour-free air is then drawn. The observer holds the syringe to his nostrils and drives the mixture into his nasal passages by means of plunger. Hence, the observer can determine the amount of dilution necessary to bring the odour to a barely perceptible level. Concentration is expressed in terms of odour unit, which is defined as the amount of odour necessary to contaminate 1 cu ft. (0.027 $m^3$) of air to the threshold level.

Odorants in the atmosphere or gas streams may also be collected by passing known volumes of air or gas through a column of activated carbon or by condensation techniques. The collected odorants may then be taken to a laboratory and released by vacuum desorption and distillation, respectively.

**Odour control**

An important problem of air pollution is the control of objectionable odours, which may be gases, mists, or solids discharged into the atmosphere from industrial, commercial, and municipal operations. Source control is the most effective means of abating odour. In many cases, this can be achieved by good sanitation practices, as the most persistent and offensive odours arise due to putrefaction.

Following are the methods which may be employed alone or in various combinations to eliminate or diminish odours.

1. Modification of the process
2. Dilution by ventilation or dispersal
3. Absorption
4. Adsorption
5. Combustion or oxidation
   (a) Direct incineration
   (b) Catalytic incineration
6. Odour masking
7. Odour counteraction of neutralization
8. Injection of a reactive substance
9. Irradiation.

**Modification of the process**

In some cases a change in the process, either by way in the composition of process materials or removal of impurities may help in odour control. Methods include substitution of low-odour solvents for highly odourous ones, adjustment of process variables like temperature, residence time etc. If such alteration makes the resultant source less intense or more tolerable from the point of odour and if such modification is technically and economically feasible, then it, merits first consideration.

**Dilution by ventilation or dispersal**

Odour intensity being a function of the odorant concentration, proper well designed ventilation is the most common method for removal of odours from enclosed spaces.

A method sometimes used for odour abatement outdoors is to release odourous gases from tall stacks. It results in normal dispersion in the atmosphere and consequent decrease in ground-level concentrations below the threshold value. Dispersal by stacks requires careful consideration of the location, meteorological parameters, etc.

**Absorption**

Absorption is applicable when the odourous gases are soluble or emulsifiable in a liquid or react chemically in solution. Liquid scrubbing of the gases in a suitable absorption unit is an important methods of odour control.

**Adsorption**

Adsorption, particularly on activated carbon, has been widely used in odour control because activated carbon has a preferential attraction and high retentivity for organic vapours. Activated carbon is highly porous and has great adsorptive power due to very large surface area. The highly porous structure permits the carbon to remove and hold the organic vapours, hydrogen sulphide, and other odour-producing substances. Furthermore, the retained material may be disrobed comparatively easily, and the carbon reactivates and is used again. But, if the concentration of the odourous material is high, the method may not be economical.

**Combustion or oxidation**

***Direct incineration***

In this process, the odour generating gases are made to pass through a combustion chamber at a temperature of the order of 650-815°C, in the presence of excess oxygen. The main demerit of this method is the cost of the fuel to produce the required temperature. Sometimes heat exchangers may be employed to recover the heat from the hot incinerated gases. The heat recovered may be used for other purposes like preheating the effluent gases, generating steam etc. The optimum residence time of the gases in the chamber should be found out by trial, since the various gas constituent to be burnt react at different rates. Normally this residence time should not be less than 0.3 sec.

**Catalytic incineration**

When the cost of heating the gas stream to 650°C is very high, catalytic combustion may be the choice. In a catalytic unit, oxidation takes place at a much lower temperature than necessary for direct incineration. In this method, the process gases, pass through specially designed units containing catalyst elements, on the surface of which oxidation occurs. During the catalytic oxidation, the constituents in the gas stream such as hydrocarbons and other organic malodours, react with oxygen to form carbon dioxide and water vapour. In general, complete oxidation should take place to solve the odour problem. Catalytic combustion of most of the organic constituents that occur in the effluent as gas stream is initiated and self sustaining at about 260°C. In case the gas stream is colder than this before coming into contact with the catalytic, it must be pre-heated. However, the fuel cost here for pre-heating will be very much less, when compared to direct incineration.

Usually certain metals and metal-oxides are used as catalysts in oxidation reactions. It has been reported that platinum and platinum-rhodium alloy distributed on a suitable support are highly effective as catalysts.

The important properties for a commercial catalyst unit are high specific surface, low resistance to the flow of gases, and arrangements in a manner as to provide turbulent mixing of process gases. Gases entering the unit should be free from particulate mater such as dust particles that might cause catalyst loss due to abrasion action. The gas stream should also be free from fumes of mercury, lead and zinc, that might impair catalyst activity by coating the catalyst or deactivating its active centres. All necessary precautions must be taken to prevent the catalyst from becoming ineffective.

Commercial catalyst units are being used for the odour control of effluent gases from chemical plants, varnish manufacture, paint and enamel-backing ovens, refuse incinerators, refineries, burning waste cracking gases and coffee roasters.

**Odour masking**

Odour masking is based on the principle that, when two odours are mixed, the stronger one will predominate. In other words, strong odours tend to mask weaker ones. Thus a strong, pleasant odour can be used to 'mask' or 'cover' weak unpleasant odour. However, care must be taken to see that the odorant used for superimposing the pleasant odour is not flammable, corrosive, or allergic.

Following are the methods used for odour masking:

1. Spraying, vaporizing, or atomizing the odorant chosen, into air-gas streams in stacks.
2. Adding directly to a process wherever possible.
3. Adding to scrubbing liquors.

In cases where the masking odorants are directly added to the process, care should be taken to see that they will not affect the quality of the products.

Odour masking can also be used to control odours in outdoor places like refuse dumps or waste lagoons. Here, the masking compound must vaporize rapidly enough to overcome the unpleasant odour and slow enough to last for a reasonable length of time.

Due to various practical problems involved on odour masking the selection of the suitable odorant and its quantity should be left to specialists in the field.

**Odour counteraction**

Certain pairs of odours of relative concentrations are antagonistic. Therefore, when they are mixed together, the intensity of each odour is diminished. This effect is known as counteraction or neutralization. Examples of such odour pairs are musk and bitter almond; rubber and cedar wood.

Selection of the proper counteractant for a given

odour is even more difficult than the selection of masking compounds, and therefore the selection should be left to experts.

**Control of air pollution by process changes**

In controlling air pollution by process changes, the method to be employed naturally depends upon the particular process involved. As a result, no fixed set of rules can be applied on a universal basis.

Generally, four methods are available to control the pollutants by process changes. They are:

1. Substitution of raw materials or fuels
2. Modification of the process itself
3. Modification or replacement of the process equipment
4. Changes in operational practices

**Substitution of raw materials or fuels**

This method has been used successfully in many cases for controlling atmospheric pollution. For example use of low-volatile coals in place of high-volatile coals has proved quite effective in eliminating smoke and soot in many industrial and commercial heating applications. Similarly, substitution of low sulphur fuels for high sulphur fuels has reduced considerably the sulphur dioxide discharge into the atmosphere. Another method of decreasing emission of air pollution is substituting bauxite flux for fluorspar in an open health furnace.

**Modification of the process**

While modifying a given process, a unit operation may be eliminated or altered, or other unit operations may be substituted or added. For example, in disposing combustible refuse, the practice of incineration may be discontinued in favour of sanitary land fill. Another example is in Brass-Foundry Practice, where an additional operational step has been used to reduce air pollution. Here, a fluxing material is applied to the surface of the molten brass which serves as an evaporation barrier and consequently reduces the emission of brass fumes. This additional step has been included strictly as an air pollution control measure. A third example of process modification is substitution of oxygen for air in gas manufacture and in blast furnaces. The vent gases produced as a result are lower in volume

**Modification or replacement of the process equipment**

This may include (a) modification of one or more items of the process equipment, (b) replacement or repair of faulty or repair of faulty or malfunctioning equipment, or (c) substitution of one type of equipment for another type.

For example

1. Use of vapour recovery systems to control vapour losses in handling volatile materials-absorbers, condensers, and compressors may bc used.
2. In cast iron founders, substitution of reverberatory furnaces for cupolas has resulted in reducing atmospheric pollution from this type of operation.

**Changes in operational practices**

A good example for this is a thermal power plant. Here by using a low-sulphur fuel in place of high-sulphur fuel, the rate of release of sulphur dioxide from coal burning operations can be reduced, especially during the periods of adverse meteorological conditions, and consequently reduce the air pollution problem.

Investigations have shown that fuel additives are effective in reducing smoke gas turbines. Hence, probably they may be used to reduce emission of smoke, soot, carbon monoxide, and hydrocarbons, associated with incomplete combustion, and thus reducing the air pollution problem.

# 8

# Air Pollution Act, 1981

An Act to provide for the prevention, control and abatement of air pollution, for the establishment, with a view to carrying out the aforesaid purposes, of Boards, for conferring on and assigning to such Boards powers and functions relating thereto and for matters connected therewith.

Whereas decisions were taken at the United Nations Conference on the Human Environment held in Stockholm in June, 1972, in which India participated, to take appropriate steps for the preservation of the natural resources of the earth which, among other things, include the preservation of the quality of air and control of air pollution;

And Whereas it is considered necessary to implement the decisions aforesaid in so far as they relate to the preservation of the quality of air and control of air pollution;

Be it enacted by Parliament in the Thirty-second Year of the Republic of India as follows:-

## CHAPTER I
### Preliminary

**Short title, extent and commencement**
1.(1) This Act may be called the Air (Prevention and Control of Pollution) Act, 1981.

(2) It extends to the whole of India.

(3) It shall come into force on such date as the Central Government may, by notification in the Official Gazette, appoint.

**Definitions**

2. In this Act, unless the context otherwise requires,-

(a) "air pollutant" means any solid, liquid or gaseous substance (including noise) present in the atmosphere in such concentration as may be or tend to be injurious to human beings or other living creatures or plants or property or environment;

(b) "air pollution" means the presence in the atmosphere of any air pollutant;

(c) "approved appliances" means any equipment or gadget used for the bringing of any combustible material or for generating or consuming any fume, gas of particulate matter and approved by the State Board for the purpose of this Act;

(d) "approved fuel" means any fuel approved by the State Board for the purposes of this Act;

(e) "automobile" means any vehicle powered either by internal combustion engine or by any method of generating power to drive such vehicle by burning fuel;

(f) "Board" means the Central Board or State Board;

(g) "Central Board" means the (Central Board for the Prevention and Control of Water Pollution) constituted under section 3 of the Water (Prevention and Control of Pollution) Act, 1974;

(h) "chimney" includes any structure with an opening or outlet from or through which any air pollutant may be emitted;

(i) "control equipment' means any apparatus, device, equipment or system to control the quality and manner of emission of any air pollutant and includes any device used for securing the efficient operation of any industrial plant;

(j) "emission" means any solid or liquid or gaseous substance coming out of any chimney, duct or flue or any other outlet;

(k) "industrial plant" means any plant used for any industrial or trade purposes and emitting any air pollutant into the atmosphere;

(l) "member" means a member of the Central Board or a State Board, as the case may be, and includes the Chairman thereof;

(m) "occupier," in relation to any factory or premises, means the person who has control over the affairs of the factory or the premises, and includes, in relation to any substance, the person in possession of the substance;

(n) "prescribed" means prescribed by rules made under this Act by the Central Government or as the case may be, the State Government;

(o) "State Board" means,-

(i) In relation to a State in which the Water (Prevention and Control of Pollution) Act, 1974, is in force and the State Government has constituted for that State a (State Board for the Prevention and Control of Water Pollution) under section 4 of that Act, the said State Board; and

(ii) in relation to any other State, the State Board for the Prevention and Control of Air Pollution constituted by the State Government under section 5 of this Act.

## CHAPTER II

## Central and State Boards for the Prevention and Control of Air Pollution

**Central Board for the Prevention and Control of Air Pollution**

3. The Central Board for the Prevention and Control of Water Pollution constituted under section 3 of the Water (Prevention and Control of Pollution) Act, 1974 (6 of 1974), shall, without prejudice to the exercise and performance of its powers and functions under this Act, exercise the powers and perform the functions of the Central Board for the Prevention and Control of Air Pollution under this Act.

**State Boards for the Prevention and Control of Water Pollution to be State Boards for the Prevention and Control of Air Pollution**

4. In any State in which the Water (Prevention and Control of Pollution) Act, 1974 (6 of 1974), is in force and the State Government has constituted for that State a State Board for the Prevention and Control of Water Pollution under section 4 of that Act, such State Board shall be deemed to be the State Board for the Prevention and Control of Air Pollution constituted under section 5 of this Act and accordingly that State Board for the Prevention and Control of Water Pollution shall, without prejudice to the exercise and performance of its powers and functions under that Act, exercise the powers and perform the functions of the State Board for the Prevention and Control of Air Pollution under this Act.

**Constitution of State Boards**

5.(1) In any State in which the Water (Prevention and Control of Pollution) Act, 1974 (6 of 1974), is not in force, or that Act is in force but the State Government has not constituted a (State Board for the Prevention and Control of Water Pollution) under that Act, the State Government shall, with effect from such date as it may, by notification in the Official Gazette, appoint, constitute a State Board for the Prevention and Control of Air Pollution under such name as may be specified in the notification, to exercise the powers conferred on, and perform the functions assigned to, that Board under this Act.

2. A State Board constituted under this Act shall consist of the following members, namely:-

(a) a Chairman, being a person, having a person having special knowledge or practical experience in respect of matters relating to environmental protection, to be nominated by the State Government:

Provided that the Chairman may be either whole-time or part-time as the State Government may think fit;

(b) Such number of officials, not exceeding five, as the State Government may think fit, to be nominated by the State Government to represent that government;

(c) such number of persons, not exceeding five, as the State Government may think fit, to be nominated by the State Government from amongst the members of the local authorities functioning within the State;

(d) such number of non-officials, not exceeding three, as the State Government may think fit, to be nominated by the State Government to represent the interest of agriculture, fishery or industry or trade or labour or any other interest, which in the opinion of that government, ought to be represented;

(e) two persons to represent the companies or corporations owned, controlled or managed by the State Government, to be nominated by that Government;

(f) a full-time member-secretary having such qualifications knowledge and experience of scientific, engineering or management aspects of pollution control as may be prescribed, to be appointed by the State Government.

Provided that the State Government shall ensure that not less than two of the members are persons having special knowledge or practical experience in respect of matters relating to the improvement of the quality of air or the prevention, control or abatement of air pollution.

(3) Every State Board constituted under this Act shall be a body corporate with the name specified by the State Government in the notification issued under sub-section (1), having perpetual succession and a common seal with power, subject to the provisions of this Act, to acquire and dispose of property and to contract, and may be the said name sue or be sued.

**Central Board to exercise the powers and perform the functions of a State Board in the Union territories**

6. No State Board shall be constituted for a Union territory and in relation to a Union territory, the Central Board shall exercise the powers and perform the functions of a State Board under this Act for that Union territory:

Provided that in relation to any Union territory the Central Board may delegate all or any of its powers and functions under this section to such person or body of persons as the Central Government may specify.

**Terms and conditions of service of members**

7.(1) Save as otherwise provided by or under this Act, a member of a State Board constituted under this Act, other than the member-secretary, shall hold office for a term of three years from the date on which his nomination is notified in the Official Gazette:

Provided that a member shall, notwithstanding the expiration of his term, continuc to hold office until his successor enters upon his office.

(2) The terms of office of a member of a State Board constituted under this Act and nominated under clause (b) or clause (e) of sub-section (2) of section 5 shall come to an end as soon as he ceases to hold the office under the State Government as the case may be, the company or corporation owned, controlled or managed by the State Government, by virtue of which he was nominated.

(3) A member of a State Board constituted under this Act, other than the member-secretary, may at any time resign his office by writing under his hand addressed,-

(a) in the case of the Chairman, to the State Government; and

(b) in any other case, to the Chairman of the State Board, and the seat of the Chairman or such other member shall thereupon become vacant.

(4) A member of a State Board constituted under this Act, other than the member-secretary, shall be deemed to have vacated his seat, if he is absent without reason, sufficient in the opinion of the State Board, from three consecutive meetings of the State Board or where he is nominated under clause (c) of sub-section (2) of section 5, he ceases to be a member of the local authority and such vacation of seat shall, in either case, take effect from such as the State Government may, by notification in the Official Gazette, specify.

(5) A casual vacancy in a State Board constituted under this Act shall be filled by a fresh nomination and the person nominated to fill the vacancy shall hold office only for the remainder of the term for which the member whose place he takes was nominated.

(6) A member of a State Board constituted under this Act shall be eligible for re-nomination

(7) The other terms and conditions of service of the Chairman and other members (except the member-secretary) of a State Board constituted under this Act shall be such as may be prescribed.

**Disqualifications**

8.(1) No person shall be a member of a State Board constituted under this Act, who-

(a) is, or at any time has been, adjudged insolvent, or

(b) is of unsound mind and has been so declared by a competent court, or

(c) is, or has been, convicted of an offence which, in the opinion of the State Government, involves moral turpitude, or

(d) is, or at any time has been, convicted of an offence under this Act, or

(e) has directly or indirectly by himself on by any partner, any share or interest in any firm or company carrying on the business of manufacture, sale, or hire of machinery, industrial plant, control equipment or any other apparatus for the improvement of the quality of air or for the prevention, control or abatement of air pollution, or

(f) is a director or a secretary, manager or other salaried officer or employee of any company or firm having any contract with the Board, or with the Government constituting the Board or with a local authority in the State, or with a company or corporation owned, controlled or managed by the Government, for the carrying out of programmes for the improvement of the quality of air or for the prevention, control or abatement of air pollution, or

(g) has so abused, in the opinion of the State Government, his position as a member, as to render his continuance on the State Board detrimental to the interest of the general public.

(2) The State Government shall, by order in writing, remove any member who is, or has become, subject

to any disqualification mentioned in sub-section (1).

Provided that no order of removal shall be made by the State Government under this section unless the member concerned has been given a reasonable opportunity of showing cause against the same.

(3) Notwithstanding anything contained in sub-section (1) or sub-section (6) of section 7, a member who has been removed under this section shall not be eligible to continue to hold office until his successor enters upon his office, or, as the case may be, for re-nomination as a member.

**Vacation of seats by members**

9. If a member of a State Board constituted under this Act becomes subject to any of the disqualifications specified in section 8, his seat shall become vacant.

**Meeting of Board**

10.(1) For the purposes of this Act, a Board shall meet at least once in every three months and shall observe such rules of procedure in regard to the transaction of business at its meetings as may be prescribed:

Provided that if, in the opinion of the Chairman, any business of an urgent nature is to be transacted, he may convene a meeting of the Board at such time as he thinks fit for the aforesaid purpose.

(2) Copies of minutes of the meetings under sub-section (1) shall be forwarded to the State Board and to the State Government concerned.

**Constitution of committees**

11.(1) A Board may constitute as many committees

consisting wholly of members or partly of members and partly of other persons and for such purpose or purposes as it may think fit.

(2) A committee constituted under this section shall meet at such time and at such place, and shall observe such rules of procedure in regard to the transaction of business at its meetings, as may be prescribed.

(3) The members of a committee other than the members of the Board shall be paid such fees and allowances, for attending its meetings and for attending to any other work of the Board as may be prescribed.

**Temporary association of persons with Board for particular purposes**

12 (1) A Board may associate with itself in such manner, and for such purposes, as may be prescribed, any person whose assistance or advice it may desire to obtain in performing any of its functions under this Act.

(2) A person associated with the Board under sub-section (1) for any purpose shall have a right to take part in the discussions of the Board relevant to that purpose, but shall not have a right to vote at a meetings of the Board and shall not be a member of the Board for any other purpose.

(3) A person associated with a Board under sub-section (1) shall be entitled to receive such fees and allowances as may be prescribed.

**Vacancy in Board not to invalidate acts or proceedings**

13. No act or proceeding of a Board or any committee thereof shall be called in question on the ground

merely of the existence of any vacancy in or any defect in the constitution of, the Board or such committee, as the case may be.

**Member-secretary and officers and other employees of State Boards**

14.(1) The terms and conditions of service of the member-secretary of a State Board constituted under this Act shall be such as may be prescribed.

(2) The member-secretary of a State Board, whether constituted under this Act or not, shall exercise such powers and perform such duties as may be prescribed or as may, from time to time, be delegated to him by the State Board or its Chairman.

(3) Subject to such rules as may be made by the State Government in this behalf, a State Board, whether constituted under this Act or not, may appoint such officers and other employees as it considers necessary for the efficient performance of its functions under this Act.

(4) The method of appointment, the conditions of service and the scale of pay of the officers (other than the member-secretary) and other employees of a State Board appointed under sub-section(3) shall be such as may be determined by regulations made by the State Board under this Act.

(5) Subject to such conditions as may be prescribed, a State Board constituted under this Act may from time to time appoint any qualified person to be a consultant to the Board and pay him such salary and allowances or fees, as it thinks fit.

**Delegation of powers**

15. A State Board may, by general or special order,

delegate to the Chairman or the member-secretary or any other officer of the Board subject to such conditions and limitations, if any, as may be specified in the order, such of its powers and functions under this Act as it may deem necessary.

## CHAPTER III
## Powers and Functions of Boards

### Functions of Central Board

16.(1) Subject to the provisions of this Act, and without prejudice to the performance, of its functions under the Water(Prevention and Control of Pollution) Act, 1974 (6 of 1974), the main functions of the Central Board shall be to improve the quality of air and to prevent, control or ovate air pollution in the country.

(2) In particular and without prejudice to the generality of the foregoing functions, the Central Board may-

(a) advise the Central Government on any matter concerning the improvement of the quality of air and the prevention, control or abatement of air pollution;

(b) plan and cause to be executed a nation-wide programme for the prevention, control or abatement of air pollution;

(c) co-ordinate the activities of the State and resolve disputes among them;

(d) provide technical assistance and guidance to the State Boards, carry out and sponsor investigations and research relating to problems of air pollution and prevention, control or abatement of air pollution;

(d) perform such function of any State Board as may be specified in and order made under sub-section (2) of section 18;

(e) plan and organise the training of persons engaged or to be engaged in programmes for the prevention, control or abatement of air pollution on such terms and conditions as the Central Board may specify;

(f) organise through mass media a comprehensive programme regarding the prevention, control or abatement of air pollution;

(g) Collect, compile and publish technical and statistical data relating to air pollution and the measures devised for its effective prevention, control or abatement and prepare manuals, codes or guides relating to prevention, control or abatement of air pollution;

(h) lay down standards for the quality of air;

(i) collect and disseminate information in respect of matters relating to air pollution;

(j) perform such other functions as may be prescribed.

(3) The Central Board may establish or recognise a laboratory or laboratories to enable the Central Board to perform its functions under this section efficiently.

(4) The Central Board may-

(a) delegate any of its functions under this Act generally or specially to any of the committees appointed by it;

(b) do such other things and perform such other acts as it may think necessary for the proper discharge of its functions and generally for the purpose of carrying into effect the purposes of this Act.

**Functions of State Boards**

17.(1) Subject to the provisions of this Act, and without prejudice to the performance of its functions, if any, under the Water (Prevention and Control of Pollution) Act, 1974 (Act 6 of 1974), the functions of a State Board shall be-

(a) to plan a comprehensive programme for the prevention, control of abatement of air pollution and to secure the execution thereof;

(b) to advise the State Government on any matter concerning the prevention, control or abatement of air pollution;

(c) to collect and disseminate information relating to air pollution;

(d) to collaborate with the Central Board in organising the training of persons engaged or to be engaged in programmes relating to prevention, control or abatement of air pollution and to organise mass-education programme relating thereto;

(e) to inspect, at all reasonable times, any control equipment, industrial plant or manufacturing process and to give, by order, such directions to such persons as it may consider necessary to take steps for the prevention, control or abatement of air pollution;

(f) to inspect air pollution control areas at such intervals as it may think necessary, assess the quality of air therein and take steps for the prevention, control or abatement of air pollution in such areas;

(g) to lay down, in consultation with the Central Board and having regard to the standards for the quality of air laid down by the Central Board, standards for emission of air pollutants into the atmosphere from industrial plants and automobiles or for the discharge of any air pollutant into the atmosphere from any other source whatsoever not being a ship or an aircraft:

Provided that different standards for emission may be laid down under this clause for different industrial plants having regard to the quantity and composition of emission of air pollutants into the atmosphere from such industrial plants;

(h) to advise the State Government with respect to the suitability of any premises or location for carrying on any industry which is likely to cause air pollution;

(i) to perform such other functions as may be prescribed or as may, from time to time, be entrusted to it by the Central Board or the State Government;

(j) to do such other things and to perform such other acts as it may think necessary for the proper discharge of its functions and generally for the purpose of carrying into effect the purposes of this Act.

(2) A State Board may establish or recognise a laboratory or laboratories to enable the State Board to perform its functions under this section efficiently.

**Power to give directions**

18.(1) In the performance of its functions under this Act-

(a) the Central Board shall be bound by such directions in writing as the Central Government may give to it; and

(b) every State Board shall be bound by such directions in writing as the Central Board or the State Government may give to it;

Provided that where a direction given by the State Government is inconsistent with the direction given by the Central Board, the matter shall be referred to the Central Government for its decision.

(2) Where the Central Government is of the opinion that any State Board has defaulted in complying with any directions given by the Central Board under sub-section (1) and as a result of such default a grave emergency has arisen and it is necessary or expedient so to do in the public interest, it may, by order, direct the Central Board to perform any of the functions of the State Board in relation to such area, for such period and for such purposes, as may be specified in the order.

(3) Where the Central Board performs any of the functions of the State Board in pursuance of a direction under sub-section (2), the expenses, if any incurred by the Central Board with respect to the performance of such functions may, if the State Board is empowered to recover such

expenses, be recovered by the Central Board with interest (at such reasonable rate as the Central Government may, by order, fix) from the date when a demand for such expenses is made until it is paid from the person or persons concerned as arrears of land revenue or of public demand.

(4) For the removal of doubts, it is hereby declared that any directions to perform the functions of any State Board given under sub-section (2) in respect of any area would not preclude the State Board from performing such functions in any other area in the State or any of its other functions in that area.

## CHAPTER IV
## Prevention and Control of Air Pollution

**Power to declare air pollution control areas**

19.(1) The State Government may, after consultation with the State Board, by notification in the Official Gazette declare in such manner as may be prescribed, any area or areas within the State as air pollution control area or areas for the purposes of this Act.

(2) The State Government may, after consultation with the State Board, by notification in the Official Gazette,-

(a) alter any air pollution control area whether by way of extension or reduction;

(b) declare a new air pollution control area in which may be merged one or more existing air pollution control areas or any part or parts thereof.

(3) If the State Government, after consultation with the State Board, is of opinion that the use of any

fuel, other than an approved fuel, in any air pollution control area or part thereof, may cause or is likely to cause air pollution, it may by notification in the Official Gazette, prohibit the use of such fuel in such area or part thereof with effect from such date (being not less than three months from the date of publication of the notification) as may be specified in the notification.

(4) The State Government may, after consultation with the State Board, by notification in the Official Gazette, direct that with effect from such date as may be specified therein, no appliance, other than an approved appliance, shall be used in the premises situated in an air pollution control area:

Provided that different dates may be specified for different parts of an air pollution control area or for the use of different appliances.

(5) If the State Government, after consultation with the State Board, is of opinion that the burning of any material (not being fuel) in any air pollution control area or part thereof may cause or is likely to cause air pollution, it may, by notification in the Official Gazette, prohibit the burning of such material in such area or part thereof.

**Powers to give instructions for ensuring standards for emission from automobiles**

20. With a view to ensuring that the standards for emission of air pollutants from automobiles laid down by the State Board under clause (g) of sub-section (1) of section 17 are complied with the State Government shall, in consultation with, the State Board, give such instructions as may be deemed necessary to the concerned authority in charge of registration of motor vehicles under the

Motor Vehicles Act, 1939 (Act 4 of 1939), and such authority shall, notwithstanding anything contained in that Act or the rules made thereunder be bound to comply with such instructions.

**Restrictions on use of certain industrial plants**

21.(1) Subject to the provisions of this section, no person shall, without the previous consent of the State Board, establish or operate any industrial plant in an air pollution control area:

**47 of 1987**

Provided that a person operating any industrial plant in any air pollution control area immediately before the commencement of section 9 of the Air (Prevention and Control of Pollution) Amendment Act, 1987, for which no consent was necessary prior to such commencement, may continue to do so for a period of three months from such commencement or, if he has made an application for such consent within the said period of three months, till the disposal of such application.

(2) An application for consent of the State Board under sub-section(1) shall be accompanied by such fees as may be prescribed and shall be made in the prescribed form and shall contain the particulars of the industrial plant and such other particulars as may be prescribed:

Provided that where any person, immediately before the declaration of any area as an air pollution control area, operates in such area any industrial plant, such person shall make the application under this sub-section within such period (being not less than three months from the date of such declaration) as may be prescribed and where such person makes such application, he shall be deemed to be operating such industrial plant with the consent of the State Board until the consent applied for has been refused.

(3) The State Board may make such inquiry as it may deem fit in respect of the application for consent referred to in sub-section (1) and in making any such inquiry, shall follow such procedure as may be prescribed.

(4) Within a period of four months after the receipt of the application for consent referred to in sub-section (1), the State Board shall, by order in writing, (and for reasons to be recorded in the order, grant the consent applied for subject to such conditions and for such period as may be specified in the order, or refuse consent.)

Provided that it shall be open to the State Board to cancel such consent before the expiry of the period for which it is granted or refuse further consent after such expiry if the conditions subject to which such consent has been granted are not fulfilled:

Provided further that before cancelling a consent or refusing a further consent under the first provision, a reasonable opportunity of being heard shall be given to the person concerned.

(5) Every person to whom consent has been granted by the State Board under sub-section (4), shall comply with the following conditions, namely:-

(i) the control equipment of such specifications as the State Board may approve in this behalf shall be installed and operated in the premises where the industry is carried on or proposed to be carried on;

(ii) the existing control equipment, if any, shall be altered or replaced in accordance with the directions of the State Board;

(iii) the control equipment referred to in clause (i) or clause (ii) shall be kept at all times in good running condition;

(iv) chimney, wherever necessary, of such specifications as the State Board may approve in this behalf shall be erected or re-erected in such premises;

(v) such other conditions as the State Board, may specify in this behalf; and

(vi) the conditions referred to in clauses (i), (ii) and (iv) shall be complied with within such period as the State Board may specify in this behalf:

Provided that in the case of a person operating any industrial plant in an air pollution control area immediately before the date of declaration of such area as an air pollution control area, the period so specified shall not be less than six months:

**Provided further that-**

(a) after the installation of any control equipment in accordance with the specifications under clause (i), or

(b) after the alteration or replacement of any control equipment in accordance with the directions of the State Board under clause (ii), or

(c) after the erection or re-erection of any chimney under clause (iv), no control equipment or chimney shall be altered or replaced or, as the case may be, erected or re-erected except with the previous approval of the State Board.

(6) If due to any technological improvement or otherwise the State Board is of opinion that all or any of the conditions referred to in sub section (5) require or requires variation (including the change of any control equipment, either in whole or in part), the State Board shall, after giving the person to whom consent has been granted an opportunity of being heard, vary all or any of such conditions and thereupon such person shall be bound to comply with the conditions as so varied.

(7) Where a person to whom consent has been granted by the State Board under sub-section (4) transfers his interest in the industry to any other person, such consent shall be deemed to have been granted to such other person and he shall be bound to comply with all the conditions subject to which it was granted as if the consent was granted to him originally.

**Persons carrying on industry, etc., not to allow emission of air pollutants in excess of the standards laid down by State Board**

22. No person operating any industrial plant, in any air pollution control area shall discharge or cause or permit to be discharged the emission of any air pollutant in excess of the standards laid down by the State Board under clause (g) of sub-section (1) of section 17.

**Power of Board to make application to court for restraining persons from causing air pollution**

22.A.(1) Where it is apprehended by a Board that emission of any air pollutant, in excess of the standards laid down by the State Board under clause (g) of sub-section (1) of section 17, is likely to occur by reason of any person operating an industrial plant or otherwise in any air pollution control area, the Board may make an application to a court, not inferior to that of a Metropolitan Magistrate or a Judicial Magistrate of the first class for restraining such person from emitting such air pollutant.

(2) On receipt of the application under sub-section (1), the court may make such order as it deems fit.

(3) Where under sub-section (2), the court makes an order restraining any person from discharging or causing or permitting to be discharged the emission of any air pollutant, it may, in that order,-

(a) direct such person to desist from taking such action as it likely to cause emission;

(b) authorize the Board, if the direction under clause (a) is not complied with by the person to whom such direction is issued, to implement the direction in such manner as may be specified by the court.

(4) All expenses incurred by the Board in implementing the directions of the court under clause (b) of sub-section (3) shall be recoverable from the person concerned as arrears of land revenue or of public demand.

**Furnishing of information to State Board and other agencies in certain cases**

23.(1) Where in any area the emission of any air pollutant into the atmosphere in excess of the standards laid down by the State Board occurs or is apprehended to occur due to accident or other unforeseen act or event, the person in charge of the premises from where which emission occurs or is apprehended to occur shall forthwith intimate the fact of such occurrence or the apprehension of such occurrence to the State Board and to such authorities or agencies as may be prescribed.

(2) On receipt of information with respect to the fact or the apprehension of any occurrence of the nature referred to in sub-section (1), whether through intimation under that sub-section or otherwise, the State Board and the authorities or agencies shall, as early as practicable, cause such remedial measure to be taken as are necessary to mitigate the emission of such air pollutants.

(3) Expenses, if any incurred by the State Board, authority or agency with respect to the remedial measures referred to in sub-section (2) together with interest (at such reasonable rate, as the State Government may, by order, fix) from the date when a demand for the expenses is made until it is paid, may be recovered by that Board, authority or agency from the person concerned, as arrears of land revenue, or of public demand.

**Power of entry and inspection**

24.(1) Subject to the provisions of this section, any person empowered by a State Board in this behalf shall have a right to enter, at all reasonable times with such assistance as he considers necessary, any place-

(a) for the purpose of performing any of the functions of the State Board entrusted to him:

(b) for the purpose of determining whether

and if so in what manner, any such functions are to be performed or whether any provisions of this Act or the rules made thereunder or any notice, order, direction or authorization served, made, given or granted under this Act is being or has been complied with;

(c) for the purpose of examining and testing any control equipment, industrial plant, record, register, document or any other material object or for conducting a search of any place in which he has reason to believe that an offence under this Act or the rules made thereunder has been or is being or is about to be committed and for seizing any such control equipment, industrial plant, record, register, document or other material object if he has reasons to believe that it may furnish evidence of the commission of an offence punishable under this Act or the rules made thereunder.

(2) Every person operating any control equipment or any industrial plant, in an air pollution control area shall be bound to render all assistance to the person empowered by the State Board under sub-section (1) for carrying out the functions under that sub-section and if he fails to do so without any reasonable cause or excuse, he shall be guilty of an offence under this Act.

(3) If any person wilfully delays or obstructs any person empowered by the State Board under sub-section (1) in the discharge of his duties, he shall be guilty of an offence under this Act.

**2 of 1974**

(4) The provisions of the Code of Criminal Procedure, 1973, or, in relation to the State of Jammu and Kashmir, or any area in which that Code is not in force, the provisions of any corresponding law in force in that State or area, shall, so far as may be, apply to any search or seizure under this section as they apply to any search or seizure made under the authority of a warrant issued under section 94 of the said Code or, as the case may be, under the corresponding provisions of the said law.

**Power to obtain information**

25. For the purposes of carrying out the functions entrusted to it, the State Board or any officer empowered by it in this behalf may call for any information (including information regarding the types of air pollutants emitted into the atmosphere and the level of the emission of such air pollutants) from the occupier or any other person carrying on any industry or operating any control equipment or industrial plant and for the purpose of verifying the correctness of such information, the State Board or such officer shall have the right to inspect the premises where such industry, control equipment or industrial plant is being carried on or operated.

**Power to take samples of air or emission and procedure to be followed in connection therewith**

26.(1) A State Board or any officer empowered by it in this behalf shall have power to take, for the purpose of analysis, samples of air or emission from any chimney, flue or duct or any other outlet in such manner as may be prescribed.

(2) The result of any analysis of a sample of emission taken under sub section (1) shall not be admissible

in evidence in any legal proceeding unless the provisions of sub-sections (3) and (4) are complied with.

(3) Subject to the provisions of sub-section (4), when a sample of emission is taken for analysis under sub-section (1), the person taking the sample shall-

(a) serve on the occupier or his agent, a notice, then and there, in such form as may be prescribed, of his intention to have it so analyzed;

(b) in the presence of the occupier or his agent, collect a sample of emission for analysis;

c) cause the sample to be placed in container or containers which shall be marked and sealed and shall also be signed both by the person taking the sample and the occupier or his agent;

(d) send, without delay, the container to the laboratory established or recognised by the State Board under section 17 or, if a request in that behalf is made by the occupier or his agent when the notice is served on him under clause (a), to the laboratory established or specified under sub-section (1) of section 28.

(4) When a sample of emission is taken for analysis under sub-section (1) and the person taking the sample serves on the occupier or his agent, a notice under clause (a) of sub-section (3), then-

(a) in a case where the occupier or his agent wilfully absents himself, the person taking

the sample shall collect the sample of emission for analysist to be placed in a container or containers which shall be marked and sealed and shall also be signed by the person taking the sample, and

(b) in a case where the occupier or his agent is present at the time of taking the sample but refuses to sign the marked and sealed container or containers of the sample of emission as required under clause (c) of subsection (3), the marked and sealed container or containers shall be signed by the person taking the sample, and the container or containers shall be sent without delay by the person taking the sample for analysis to the laboratory established or specified under sub-section (1) of section 28 and such person shall inform the Government analyst appointed under sub-section (1) of section 29, in writing, about the wilfull absence of the occupier or his agent, or, as the case may be, his refusal to sing the container or containers.

**Reports of the results of analysis on samples taken under section 26**

27.(1) Where a sample of emission has been sent for analysis to the laboratory established or recognised by the State Board, the Board analyst appointed under sub-section (2) of section 29 shall analyse the sample and submit a report in the prescribed form of such analysis in triplicate to the State Board.

(2) On receipt of the report under sub-section (1), one copy of the report shall be sent by the State Board to the occupier or his agent referred to in section 26, another copy shall be preserved for production before the court in case any legal proceedings are taken against him and the other copy shall be kept by the State Board.

(3) Where a sample has been sent for analysis under clause (d) of sub-section (3) or sub-section (4) of section 26 to any laboratory mentioned therein, the Government analyst referred to in the said sub-section (4) shall analyse the sample and submit a report in the prescribed form of the result of the analysis in triplicate to the State Board which shall comply with the provisions of sub-section (2).

(4) Any cost incurred in getting any sample analysed at the request of the occupier or his agent as provided in clause (d) of sub-section (3) of section 26 or when he wilfully absents himself or refuses to sing the marked and sealed container or containers of sample of emission under sub-section (4) of that section, shall be payable by such occupier or his agent and in case of default the same shall be recoverable from him as arrears of land revenue or of public demand.

**State Air, Laboratory**

28.(1) The State Government may, by notification in the Official Gazette,-

(a) establish one or more State Air Laboratories; or

(b) specify one or more laboratories or institutes as State Air Laboratories to carry out the functions entrusted to the State Air Laboratory under this Act.

(2) The State Government may, after consultation with the State Board, make rules prescribing-

(a) the functions of the State Air Laboratory;

(b) the procedure for the submission to the said Laboratory of samples of air or emission for analysis or tests, the form of the Laboratory's report thereon and the fees payable in respect of such report;

(c) such other matters as may be necessary or expedient to enable that Laboratory to carry out its functions.

**Analysts**

29.(1) The State Government may, by notification in the Official Gazette, appoint such persons as it thinks fit and having the prescribed qualifications to be government analysts for the purpose of analysis of samples of air or emission sent for analysis to any laboratory established or specified under sub-section (1) of section 28.

(2) Without prejudice to the provisions of section 14, the State Board may, by notification in the Official Gazette, and with the approval of the State Government, appoint such persons as it thinks fit and having the prescribed qualifications to be Board analysts for the purpose of analysis of samples of air or emission sent for analysis to any laboratory established or recognised under section 17.

**Reports of analysts**

30. Any document purporting to be a report signed by a Government analyst or, as the case may be, a State Board analyst may be used as evidence of the facts stated therein in any proceeding under this Act.

**Appeals**

31. (1) Any person aggrieved by an order made by the State Board under this Act may, within thirty days from the date on which the order is communicated to him, prefer an appeal to such authority (hereinafter referred to as the Appellate Authority) as the State government may think fit to constitute:

Provided that the Appellate Authority may entertain the appeal after the expiry of the said period of thirty days if such authority is satisfied that the appellant was prevented by sufficient cause from filing the appeal in time.

(2) The Appellate Authority shall consist of a single person or three persons as the State Government may think fit to be appoint by the State Government.

(3) The form and the manner in which an appeal may be preferred under sub-section (1), the Appellate Authority shall, after giving the appellant and the State Board an opportunity of being heard, dispose of the appeal as expeditiously as possible.

**Power to give directions**

31A. Notwithstanding anything contained in any other law, but subject to the provisions of this Act, and to any directions that the Central Government may give in this behalf, a Board may, in the exercise of its powers and performance of its functions under this Act, issue any directions in writing to any person, officer or authority, and such person, officer or authority shall be bound to comply with such directions.

Explanation. For the avoidance of doubts, it is hereby declared that the power to issue directions under this section includes the power to direct-

(a) the closure, prohibition or regulation of any industry, operation or process; or

(b) the stoppage or regulation of supply of electricity, water or any other service.]

## CHAPTER V
## Fund, Accounts and Audit

**Contributions by Central Government**

32. The Central Government may, after due appropriation made by Parliament by law in this behalf make in each financial year such contributions to the State Boards as it may think necessary to enable the State Boards as it may think necessary to enable the State Board to perform their functions under this Act:

**6 of 1974**

Provided that noting in this section shall apply to any [State Board for the Prevention and Control of water Pollution] constituted under section 4 of the Water (Prevention and Control of Pollution) Act, 1974, which is empowered by that Act to expend money from its fund thereunder also for performing its functions, under any law for the time being in force relating to the prevention, control or abatement of air pollution.

**Fund of Board**

33.(1) Every State Board shall have its own fund for the purposes of this Act and all sums which may, from time to time, be paid to it by the Central Government and all other receipts (by way of contributions, if any, from the State Government, fees, gifts, grants, donations benefactions or otherwise) of that Board shall be carried to the fund of the Board and all payments by the Board shall be made therefrom.

(2) Every State Board may expend such sums as it thinks fit for performing its functions under this Act and such sums shall be treated as expenditure payable out of the fund of that Board.

**6 of 1974**

(3) Nothing in this section shall apply to any [State Board for the Prevention and Control of Water Pollution] constituted under section 4 of the Water (Prevention and Control of Pollution) Act, 1974, which is empowered by that Act to expend money from its fund thereunder also for performing its functions under any law for the time being in force relating to the prevention, control or abatement of air pollution.

**Borrowing powers of Board**

33A. a board may, with the consent of, or in accordance with the terms of any general or special authority given to it by, the Central Government or, as the case may be, the State Government, borrow money from any source by way of loans or issue of bonds, debentures or such other instruments, as it may deem fit, for discharging all or any of its functions under this Act.]

**Budget**

34. The Central Board or as the case may be, the State Board shall, during each financial year, prepare, in such form and at such time as may be prescribed, a budget in respect of the financial year next ensuing showing the estimated receipt and expenditure under this Act, and copies thereof shall be forwarded to the Central Government or, as the case may be, the State Government.

**Annual report**

35.(1) The Central Board shall, during each financial year, prepare, in such form as may be prescribed, an annual report giving full account of its activities under this Act during the previous financial year and copies thereof shall be forwarded to the Central Government within four months from the last date of the previous financial year and that Government shall cause every such report to be laid before both Houses of Parliament within nine months of the last date of the previous financial year.

(2) Every State Board shall, during each financial year, prepare, in such form as may be prescribed, an annual report giving full account of its activities under this Act during the previous financial year and copies thereof shall be forwarded to the State Government within four months from the last date of the previous financial year and that Government shall cause every such report to be laid before the State Legislature within a period of nine months from the date of the previous financial year.]

**Accounts and audit**

36.(1) Every Board shall, in relation to its functions under this Act, maintain proper accounts and other relevant records and prepare an annual statement of accounts in such form as may be prescribed by the Central Government or, as the case may be, the State Government.

**1 of 1956**

(2) The accounts of the Board shall be audited by an auditor duly qualified to act as an auditor of companies under section 226 of the Companies Act, 1956.

(3) The said auditor shall be appointed by the Central Government or, as the case may be, the State Government on the advice of the Comptroller and Auditor General of India.

(4) Every auditor appointed to audit the accounts of the Board under this Act shall have the right to demand the production of books, accounts, connected vouchers and other documents and papers and to inspect any of the offices of the Board.

(5) Every such auditor shall send a copy of his report together with an audited copy of the accounts to the Central Government or, as the case may be, the State Government.

(6) The Central Government shall, as soon as may be after the receipt of the audit report under sub-section (5), cause the same to be laid before both Houses of Parliament.

(7) The State Government shall, as soon as may be after the receipt of the audit report under sub-section (5), cause the same to be laid before the State Legislature.

## CHAPTER VI
## Penalties and Procedure

**Failure to comply with the provisions of section 21 or section 22 or with the directions issued under section 33A**

37.(1) Whoever fails to comply with the provisions of section 21 or section 22 or directions issued under section 31A, shall, in respect of each such failure, be punishable with imprisonment for a terms which shall not be less than one year and six months but which may extends to six years with fine, and in case the failure continues, with an additional fine which may extend to five thousand rupees for every day during which such failure continues after the conviction for the first such failure.

(2) If the failure referred to in sub-section (1) continues beyond a period of one year after the date of conviction, the offender shall be punishable with imprisonment for a term which shall not be less than two years but which may extend to seven years and with fine.

**Penalties for certain acts**

38. Whoever-

(a) destroys, pulls down, removes, injures or defaces any pillar, post or stake fixed in the ground or any notice or other matter put up, inscribed or placed, by or under the authority of the Board, or

(b) obstructs any person acting under the orders or directions of the Board from exercising his powers and performing his functions under this Act, or

(c) damages any works or property belonging to the Board, or

(d) fails to furnish to the Board or any officer or other employee of the Board any information required by the Board or such officer or other employee for the purpose of this Act, or

(e) fails to intimate the occurrence of the emission of air-pollutants into the atmosphere in excess of the standards laid down by the State Board or the apprehension of such occurrence, to the State Board and other prescribed authorities or agencies as required under sub-section (1) of section 23, or

(f) in giving any information which he is required to give under this Act, makes a statement which is false in any material particular, or

(g) for the purpose of obtaining any consent under section 21, makes a statement which is false in any material particular.

Shall be punishable with imprisonment for a term which may extend to three months or with fine which may extend to (ten thousand rupees ) or with both.

**Penalty for Contravention of certain provisions of the Act**

39. Whoever contravenes any of the provisions of this Act or any order or direction issued thereunder, for which no penalty has been elsewhere provided in this Act, shall be punishable with imprisonment for a term which may extend to three months or with fine which may extend to ten thousand rupees or with both, and in the case of continuing contravention, with an additional fine which may extend to five thousand, rupees for every day during which such contravention continues after conviction for the first such contravention.

**Offences by companies**

40.(1) Where an offence under this Act has been committed by a company, every person who, at the time the offence was committed, was directly in charge of, and was responsible to, the company for the conduct of the business of the company, as well as the company, shall be deemed to be guilty of the offence and shall be liable to be proceeded against and punished accordingly:

Provided that nothing contained in this sub-section shall render any such person liable to any punishment provided in this Act, if he proves that the offence was committed without his knowledge or that he exercised all due diligence to prevent the commission of such offence.

(2) Notwithstanding anything contained in sub-section (1), where an offence under this Act has been committed by a company and it is proved that the offence has been committed with the consent or connivance of, or is attributable to any neglect on the part of, any director, manager, secretary or other officer of the company, such director, manager, secretary or other officer shall also be deemed to be guilty of that offence and shall be liable to be proceeded against and punished accordingly.

Explanation. For the purpose of this section-

(a) "company" means any body corporate, and includes a firm or other association of individuals; and

(b) "director," in relation to a firm, means a partner in the firm.

**Offences by Government Departments**

41.(1) Where an offence under this Act has been committed by any Department of Government, the Head of the Department shall be deemed to be guilty of the offence and shall be liable to be proceeded against and punished accordingly:

Provided that nothing contained in this section shall render such Head of the Department liable to any punishment if he proves that the offence was committed without his knowledge or that he exercised all due diligence to prevent the commission of such offence.

(2) Notwithstanding anything contained in sub-section (1), where an offence under this Act has been committed by a Department of Government and it is proved that the offence has been committed with the consent or connivance of, or is attributable to any neglect on the part of, any officer, other than the Head of the Department, such officer shall also be deemed to be guilty of that offence and shall be liable to be proceeded against and punished accordingly.

**Protection of action taken in good faith**

42. No suit, prosecution or other legal proceeding shall lie against the Government or any officer of the Government or any member or any officer or other employee of the Board in respect of anything which is done or intended to be done in good faith in pursuance of this Act or the rules made thereunder.

**Cognizance of offences**

43.(1) No court shall take cognizance of any offence under this Act except on a complaint made by-

(a) a Board or any officer authorized in this behalf by it; or

(b) any person who has given notice of not less than sixty days, in the manner prescribed, of the alleged offence and of his intention to make a complaint to the Board or officer authorized as aforesaid, and no court inferior to that of a Metropolitan Magistrate or a Judicial Magistrate of the first class shall try any offence punishable under this Act.

(2) Where a complaint has been made under clause (b) of sub-section (1) the Board shall, on demand by such person, make available the relevant reports in its possession to that person:

Provided that the Board may refuse to make any such report available to such person if the same is, in its opinion, against the public interest.

**Members, officers and employees of Board to be public servants**

44. All the members and all officers and other employees of a Board when acting or purporting to act in pursuance of any of the provisions of this Act or the rules made thereunder shall be deemed to be public servant within the meaning of section 21 of the Indian Penal Code (45 of 1860).

**Reports and Returns**

45. The Central Board shall, in relation to its functions under this Act, furnish to the Central Government, and a State Board shall, in relation to its functions under this Act, furnish to the State Government and to the Central Board such reports, returns, statistics, accounts and other information as that Government, or, as the case may be, the Central Board may, from time to time, require.

**Bar of Jurisdiction**

46. No civil court shall have jurisdiction to entertain any suit or proceeding in respect of any matter which an Appellate Authority constituted under

this Act is empowered by or under this Act to determine, and no injunction shall be granted by any court or other authority in respect of any action taken or to be taken in pursuance of any power conferred by or under this Act.

**Miscellaneous**

**Power of State Government to supersede State Board**

47.(1) If at any time the State Government is of opinion-

(a) that a State Board constituted under this Act has persistently made default in the performance of the functions imposed on it by or under this Act, or

(b) that circumstances exist which render it necessary in the public interest so to do, the State Government may, by notification in the Official Gazette, supersede the State Board for such period, not exceeding six months, as may be specified in the notification:

Provided that before issuing a notification under this sub-section for the reasons mentioned in clause (a), the State Government shall give a reasonable opportunity to the State Board to show cause why it should not be superseded and shall consider the explanations and objections, if any, of the State Board.

(2) Upon the publication of a notification under sub-section (1) superseding the State Board,-

(a) all the members shall, as from the date of supersession, vacate their offices as such;

(b) all the powers, functions and duties which may, by or under this Act, be exercised, performed or discharged by the State Board shall, until the State Board is reconstituted under sub-section (3), be exercised, performed or discharged by such person or persons as the State Government may direct;

(c) all property owned or controlled by the State Board shall, until the Board is reconstituted under sub-section (3), vest in the State Government.

(3) On the expiration of the period of supersession specified in the notification issued under sub-section (1), the State Government may-

(a) extend the period of supersession for such further term, not exceeding six months, as it may consider necessary; or

(b) reconstitute the State Board by a fresh nomination or appointment as the case may be, and in such case any person who vacated his office under clause (a) of sub-section (2) shall also be eligible for nomination or appointment.

Provided that the State Government may at any time before the expiration of the period of supersession whether originally specified under sub-section (1) or as extended under this sub-section, take action under clause (b) of this sub-section.

**Special provision in the case of supersession of the Central Board or the State Boards constituted under the Water (Prevention and Control of Pollution) Act, 1974**

48. Where the Central Board or any State Board constituted under the Water (Prevention and Control of Pollution) Act, 1974, is superseded by

the Central Government or the State Government, as the case may be, under that Act, all the powers, functions and duties of the Central Board or such State Board under this Act shall be exercised, performed or discharged during the period of such supersession by the person or persons, exercising, performing or discharging the powers, functions and duties of the Central Board or such State Board under the Water (Prevention and Control of Pollution) Act, 1974, during such period.

**Dissolution of State Boards constituted under the Act**

49.(1) As and when the Water (Prevention and Control of Pollution) Act, 1974 (Act 6 of 1974), comes into force in any State and the State Government constitutes a (State Board for the Prevention and Control of Water Pollution) under that Act, the State Board constituted by the State Government under this Act shall stand dissolved and the Board first-mentioned shall exercise the powers and perform the functions of the Board second-mentioned in that State.

(2) On the dissolution of the State Board constituted under this Act,-

(a) all the members shall vacate their offices as such;

(b) all moneys and other property of whatever kind (including the fund of the State Board) owned by, or vested in, the State Board, immediately before such dissolution, shall stand transferred to and vest in the State Board for the Prevention and Control of Water Pollution;

(c) every officer and other employee serving under the State Board immediately before such dissolution shall be transferred to and become an officer or other employee of the (State Board for the Prevention and Control of Water Pollution) and hold office by the same tenure and at the same remuneration and on the same terms and conditions of service as he would have held the same if the State Board constituted under this Act had not been dissolved and shall continue to do so unless and until such tenure, remuneration and conditions of service are duly altered by the (State Board for the Prevention and Control of Water Pollution):

Provided that the tenure, remuneration and terms and conditions of service of any such officer or other employee shall not be altered to his disadvantage without the previous sanction of the State Government;

(d) all liabilities obligations of the State Board of whatever kind, immediately before such dissolution, shall be deemed to be the liabilities or obligations, as the case may be, of the State Board for the Prevention and Control of Water Pollution) and any proceeding or cause of action, pending or existing immediately before such dissolution by or against the State Board constituted under this Act in relation to such liability or obligation may be continued and enforced by or against the State Board for the Prevention and Control of Water Pollution.

50. (Power to amend the Schedule) Rep. by the Air (Prevention and Control of Pollution) Amendment Act, 1987 (47 of 1987), s. 22 (w.e.f. 1-4-1988).

**Maintenance of register**

51.(1) Every State Board shall maintain a register containing particulars of the persons to whom consent has been granted under section 21, the standard for emission laid down by it in relation to each such consent and such other particulars as may be prescribed.

(2) The register maintained under sub-section (1) shall be open to inspection at all reasonable hours by any person interested in or affected by such standards for emission or by any other person authorized by such person in this behalf.

**Effect of other Laws**

52. Save as otherwise provided by or under the Atomic Energy Act, 1962 (33 of 1962), in relation to radioactive air pollution the provisions of this Act shall have effect notwithstanding anything inconsistent therewith contained in any enactment other than this Act.

**Power of Central Government to make rules**

53.(1) The Central Government may, in consultation with the Central Board by notification in the Official Gazette, make rules in respect of the following matters namely:-

(a) the intervals and the time and place at which meetings of the Central Board or any committee thereof shall be held and the procedure to be followed at such meetings, including the quorum necessary for the transaction of business thereat, under sub-section (1) of section 10 and under sub-section (2) of section 11;

(b) the fees and allowances to be paid to the members of a committee of the Central

Board, not being members of the Board, under sub-section (3) of section 11;

(c) the manner in which and the purposes for which persons may be associated with the Central Board under sub-section (1) of section 12;

(d) the fees and allowance to be paid under sub-section (3) of section 12 to persons associated with the Central Board under sub-section (1) of section 12;

(e) the functions to be performed by the Central Board under clause (j) of sub-section (2) of section 16;

(f) the form in which and the time within which the budget of the Central Board may be prepared and forwarded to the Central Government under section 34;

(f) the form in which the annual report of the Central Board may be prepared under section 35;

(g) the form in which the accounts of the Central Board may be maintained under sub-section (1) of section 36.

(2) Every rule made by the Central Government under this Act shall be laid, as soon as may be after it is made, before each House of Parliament, while it is in session, for a total period of thirty days which may be comprised in one session or in two or more successive sessions, and if, before the expiry of the session immediately following the session or the successive sessions aforesaid, both Houses agree in making any modification in the rule or both Houses agree that the rule should not be made,

the rule shall thereafter have effect only in such modified form or be of no effect, as the case may be; so, however, that any such modification or annulment shall be without prejudice to the validity of anything previously done under that rule.

Power of State Government to make rules

54.(1) Subject to the provisions of sub-section (3), the State Government may, by notification in the Official Gazette, make rules to carry out the purposes of this Act in respect of matter not falling within the purview of section 53.

(2) In particular, and without prejudice to the generality of the foregoing power, such rules may provide for all or any of the following matters, namely:-

(a) the qualifications, knowledge and experience of scientific, engineering or management aspect of pollution control required for appointment as member-secretary of a State Board constitute under the Act;

(a) the terms and conditions of service of the Chairman and other members (other than the member-secretary) of the State Board constituted under this Act under sub-section (7) of section 7;

(b) the intervals and the time and place at which meetings of the State Board or any committee thereof shall be held and the procedure to be followed at such meetings, including the quorum necessary for the transaction of business threat, under sub-section (1) of section 10 and under sub-section (2) of section 11;

(c) The fees and allowances to be paid to the members of a committee of the State Board, not being members of the Board under sub-section (3) of section 11;

(d) the manner in which and the purpose for which persons may be associated with the State Board under sub-section (1) of section 12.

(e) the fees and allowances to be paid under sub-section (3) of section 12 to persons associated with the State Board under sub-section (1) of section 12;

(f) the terms and conditions of service of the member secretary of a State Board constituted under this Act under sub-section (1) of section 14;

(g) the powers and duties to be exercised and discharged by the member-secretary of a State Board under sub-section (2) of section 14;

(h) the conditions subject to which a State Board may appoint such officers and other employees as it considers necessary for the efficient performance of its functions under sub-section (3) of section 14;

(i) the conditions subject to which a State Board may appoint a consultant under sub-section (3) of section 14;

(j) the functions to be performed by the State Board under clause (i) of sub-section (1) of section 17;

(k) the manner in which any area or areas may be declared as air pollution control area or areas under sub-section (1) of section 19;

(l) the form of application for the consent of the State Board, the fees payable therefore, the period within which such application shall be made and the particulars it may contain, under sub-section (2) of section 21;

(m) the procedure to be followed in respect of an inquiry under sub-section (3) of section 21;

(n) the authorities or agencies to whom information under sub-section (1) of section 23 shall be furnished;

(o) the manner in which samples of air or emission may be taken under sub-section (1) of section 26;

(p) the form of the notice referred to in sub-section (3) of section 26;

(q) the form of the report of the State Board analyst under sub-section (1) of section 27;

(r) the form of the report of the Government analyst under sub-section (3) of section 27;

(s) the functions of the State Air Laboratory, the procedure for the submission to the said Laboratory of samples of air or emission for analysis or tests, the form of Laboratory's report thereon, the fees payable in respect of such report and other matters as may be necessary or expedient to enable that Laboratory to carry out its functions, under sub-section (2) of section 28;

(t) the qualifications required for Government analysts under sub-section (1) of section 29;

(u) the qualification required for State Board analysts under sub-section (2) of section 29;

(v) the form and the manner in which appeals may be preferred, the fees payable in respect of such appeals and the procedure to be followed by the Appellate Authority in disposing of the appeals under sub-section (3) of section 31;

(w) the form in which and the time within which the budget of the State Board may be prepared and forwarded to the State Government under section 34;

(ww)the form in which the annual report of the State Board may be prepared under section 35;

(x) the form in which the accounts of the State Board may be maintained under the sub-section (1) of section 36;

(xx) the manner in which notice of intention to make a complaint shall be given under section 43;

(y) the particulars which the register maintained under section 51 may contain;

(z) any other matter which has to be, or may be, prescribed.

(3) After the first constitution of the State Board, no rule with respect to any of the matters referred to in sub-section (2) other than those referred to

(in clause (aa) thereof), shall be made, varied, amended or repealed without consulting that Board.

(The Schedule.) Omitted by the Air (Prevention and Control of Pollution) Amendment Act, 1987, s. 25 (w.e.f. 1-4-1988).

# 9

# Acid Rain

## Introduction

Increasing acidity in natural waters and soils is becoming a problem over extended areas of the world, particularly North Eastern America and North Western Europe. This acidity has been associated with the transport and subsequent disposition of sulphur dioxide, nitrogen oxides and their acid oxidation products. The concentration levels of the transported gases have not been sufficient to damage the environment directly, it is their accumulation which causes the problem. This accumulation was first reported as increasing acidity in Swedish lakes and rivers, an acidity that could not be explained by reference to Sweden's sulphur dioxide emissions alone. It was consequently proved that much of the material responsible got transported from the highly industrialized areas of the UK and Central Europe, and that this transport had been aided by the tall industrial stacks which had been used in these areas to avoid a local problem.

## Meaning of Acid Rains

Acid rains means in common language the presence of excessive acids in rain water. It has been one of the effects of air pollution. Every source of energy that we use—be it coal, fuelwood or petroleum products, has sulphur and nitrogen. These two elements when burnt in the presence of atmospheric oxygen are converted into their respective oxide-sulphur dioxide and nitrogen dioxide, which are highly soluble in water.

**Adverse Effects of Acid Rains**

Probably the most worrying aspect has been that the area affected by acid precipitation has been increasing year by year, and its association with our energy consuming world leaves little prospect of allevation, even with stricter emission controls.

The repercussions on the environment have been difficult to quantify because they are a function of soil type and bedrock. It does seem, however, that the silaceous bedrocks, thin soils and soft waters of the areas at present receiving the deposition have been particularly vulnerable. It gets related to the relative insolubility of the rock and the consequent lack of dissolved species in the water which could counder (buffer) the acidity change. Significant reductions in fish populations have been observed, accompanied by an associated decrease in the variety of species at all levels in the food chain. These are 15,000 fishless lakes in Sweden because of increased acidity, and some 100 such lakes in the Adirondak region of the USA. Soils in these regions similarly are having little buffering capacity and are subject to the leaching of minerals essential to plant growth. There have been soils which have been low in sulphur or high in carbonate, where an in increase in acid sulphate could well be beging or even beneficial, but these have been not predominant in North Western Europe or North Eastern USA.

Acid rain causes a number of adverse implications. It tends to increase acidity in the soil, threatens human and aquatic life, destroys, forests and crops reducing agricultural productivity. Besides, it is able to corrode buildings, monuments, statues, bridges, fences, and railings, that costs the world 1450 million dollars a year. At St. Paul's Cathedral some stone-work is being eaten away at the rate of an inch every 100 years. Even the British Parliament building has suffered serious damage from the presence of sulphuric acid in rain fall (The Times of India, October 25, 1986). It creates a serious threat to human health also, since it contaminates not only the breathing air but also the drinking water and even food.

The acids have been found to be very dangerous to the living organisms they can destroy life. Acid rain can play havoc with the human nervous system by making the person an easy prey to neurological diseases. This happens because these acids produce highly toxic compounds which contaminates the potable water and enter our body. Acid rain has been already an acute problem in North America and Europe. Crops and forest in Canada are being destroyed by acid rain due to pollutants emitted by industries in Northern USA. Acidity kills fish, bacteria and algae and the acquatic eco-system is destroyed.

Winds are known to carry air pollutants from one country to another. Air pollution in England now descends upon Sweden as acid rain. Acidification of soil changes its biology and chemistry. When the soil gets acidified, plants can absorb cadmium more easily and high levels of cadmium in plants has been dangerous for animals and human beings.

Acid rain in Japan has demanged 5,000 sq. kms of cedar trees in Kanto plan which lies to the north of Tokyo. This area is having high acid deposition brought about by air pollutants. Acidic air pollutants have been responsible for many other damaging effects like corrosion of metals, weakening or disintegration of textiles, paper and marble. Investigations are going on to know whether Taj Mahal is being affectd due to pollutants released from Mathura refinery. Hydrogen Sulphide tranishes silver and blackens leaded house paints ozone produces cracks in rubber; its economic significance has been apparent

Air pollution may cause or contribute to a variety of safety hazards e.g., hazards associated with reduced visibility due to smog etc. Air contaminants that have been detectable by the sense of sight, touch, smell or taste can be nuisance in many way even if they do not result in direct adverse economics or health effects.

In the near future the developing countries like India will soon have to cope with this problem of acid pollution. Industrial areas with the pH value of rain below or close to the critical value had been recorded in Delhi, Nagpur, Pune, Bombay and Calcutta. Nadia acidity has been largely due to sulphur dioxide from coal fired power plants and petroleum refinery, emitting nearly 85% tonnes annually. The phenomenon of acid rain is becoming more and more common in Bombay with several industries discharging sulphur oxides in the air making rain water more acidic. The average pH value of acid rain at Calcutta has been 5.80, Hyderabad, 5.73, Madras, 5.85, Delhi, 6.21, and Bombay, 4.80. The situation in India has been likely to worsen in the near future with the increased tempo of thermal power plants.

**Theory of Acid Rain**

***Oxidation of Sulphur Dioxide***

The foremost chemical reaction of $SO_2$ in the atmosphere is oxidation to $SO_3$, which with water gives sulphuric acid. The acid or sulphates, in particular $(NH_4)_2SO_4$ and $NH_4$ $HSO_4$ occur as acrosols. The process can be represented by:

$$\underset{\text{oxidation}}{SO_2 \rightarrow} SO_3 \overset{H_2O}{\rightarrow} H_2SO_4,\ (NH_4)_2SO_4,\ NH_4HSO_4 \quad \ldots(1)$$

aerosols

The oxidation step is the least well understood, but appears to proceed in one or more of three ways; catalytic oxidation or photooxidation or oxidations by rapical. In order for reactions to occur a number of environmental factors need to be favourable including the humidity, temperature, light intensity, aerosol levels and material transport within the atmosphere.

*Catalysis.* The reaction:

$$2SO_2 + 2H_2O + O_2 \rightleftharpoons 2H_2SO_4 \quad \Delta H = -600 \text{ kJ} \ldots(2)$$

is slow in clean air, but is catalysed by aerosols containing metal ions. Catalysts that increase the rate 10 to 100 times are salts of Mn (II), Fe (III) and Cu (II), and oxides of Cr (III), AI (III), Pb (II) and Ca (II).

Surfaces, such as on buildings, may also act as catalytic centres. The reaction rate is greatest in humidity >32%, and especially at high humidities = 70%. The $SO_2$ is oxidised in water droplets, in the presence of heterogeneous catalysts in the droplet. Increase in acidity inhibits the reaction, which proceeds best in neutral or alkaline solutions. Therefore since the reaction produces $H_2SO_4$ it slows down as acid accumulates. The formation of $HSO_3^-$ and $SO_3^{3-}$ (from $SO_2$ in water) are suppressed by increasing acidity driving reaction (2) to the left, reducing the solubility of $SO_2$ in the droplet, and therefore the reaction rate. The solubility of $SO_2$ in water is a function of pH and can be described by the equilibria :

$$SO_{2(g)} + H_2O \quad SO_{2(aq)} \quad \ldots(3)$$

$$SO_{2(aq)} + H_2O \quad H_2SO_{3(aq)} \quad \ldots(4)$$

$$H_2SO_3 + H_2O \quad H_3O^+ + HSO_3^- \qquad ...(5)$$

$$HSO_3^- + H_2O \quad H_3O^+ + SO_3^{2-} \qquad ...(6)$$

$$2HSO_3^- \quad S_2O_5^{2-} + H_2O \qquad ...(7)$$

The sulphur species in water are either covalent ($SO_{2(aq)}$ and $H_2SO_3$) or ionic ($HSO_3^-$, $SO_3^{2-}$ and $S_2O_5^{2-}$), the latter being more soluble because of the polarity of water. In the equilibria (5) and (6) the $H_3O^+$ occurs on the right and therefore its removal favours the ionic species. This agrees with the fact that below pH4, $SO_3$ solubility drops markedly, while it rises above pH4. Addition of ammonia promotes the reaction rate by reducing the acidity and forming $(NH_4)_2SO_4$ and $NH_4$ $HSO_4$, the main sulphate aerosols.

The reaction mechanism for equation (2) probably involves oxidation of sulphite species rather than $SO_2$ :

$$HSO_3^- + H_2O—2e \rightarrow HSO_4^- + 2H^+ \qquad ...(8)$$

$$O_2 + 4H^+ + 4e \rightarrow 2H_2O \qquad ...(9)$$

*i.e.*, $$2HSO_3^- + O_2 \rightarrow 2HSO_4^- \qquad ...(10)$$

The metal catalyst may act as an oxidising agent, for example $Cu^{2+}$ would be reduced to $Cu^+$ which is reoxidised by dioxygen. Alternatively a coordination catalyst may be involved. For example, the $Mn^{2+}$ ion could effect the oxidation:

$$Mn^{2+} + SO_2 \rightarrow [Mn—SO_2]^{2+} \qquad ...(11)$$

$$2[Mn—SO_2]^{2}+O_2 \rightarrow [Mn—SO_2]^{2+}+O_2 \rightarrow 2[Mn—SO_3]^{2}_{+} \qquad ...(12)$$

$$[Mn—SO_3]^{2+} + H_2O \rightarrow Mn^{2+} + HSO_4^- + H^+ \qquad ...(13)$$

*Photochemical oxidation*. Sulphur dioxide absorbs radiation producing excited states:

$$SO_2\ (^1A_1) + h\nu \rightarrow SO_2\ (^1A_2\ ^1B_1)$$
240—330 nm
(intense intercrossing absorption)

$$SO_2\ (^1A_1) + h\nu \rightarrow SO_2\ (^3B_1) \quad ...(15)$$
340—400 nm
(weak absorption)

The triplet state has a relatively long lifetime, and may react by a variety of routes to give $SO_3$, for example;

$$SO_3\ (^3B_1) + O_2 \rightarrow SO_3 + O\ (^3P) \quad ...(16)$$

$$SO_2\ (^3B_1) + SO_2 \rightarrow SO_3 + SO \quad ...(17)$$

$$\text{and } SO + SO_2 \rightarrow SO_3 + S \quad ...(18)$$

For levels of $SO_2$ of 5-30 ppm, reaction (16) achieves around 0.1—0.2% conversion to $SO_3$ per hour.

*Oxidation by raidcals.* Hydrocarbons and nitrogen dioxide both increase the rate of oxidation of $SO_2$, as do radicals and peroxy compounds. These materials are either components of automobile emissions of secondary products. The following reactions are feasible routes to the oxidation of $SO_2$.

$$SO_2 + O_3 \rightarrow SO_3 + O_2 \quad ...(19)$$

This reaction is slow in the gas phase, but rapid in solution (water droplet). The reaction:

$$SO_2 + NO_2 \rightarrow SO_2 + NO \quad ...(20)$$

May have a similar mechanism to the reactions producing $SO_3$ by the Lead Chamber Process. The reverse reaction is also possible in the atmosphere.

Other feasible reactions are:

$SO_2 + HO_2 \rightarrow SO_3 + HO.$ ...(21)

$SO_2 + RO_2 \rightarrow SO_3 + RO., (R = CH_3\ C = O)$ ...(22)

$HO + SO_2 \rightarrow HOSO_2.$ ...(23)

$HOSO_2 + O_2 \rightarrow HOSO_2\ G_2$ ...(24)

$HOSO_2\ O_2 + NO \rightarrow HOSO_3 + NO_2$ ...(25)

$H_2SO_4$

$RH + SO_2\ (^3B_1) \rightarrow RSO_2H \rightarrow H_2SO_4 + RH$ ...(26)

Suphinilc acid

A number of reactions are available for the oxidation of $SO_2$ to $SO_3$, which occurs extensively in the atmosphere as indicated by presence of $H_2SO_4$ aerosol.

**Acid Rain**

Acid rain is produced by the dissolving of $SO_2$ in water to give $H_2SO_3$ and subsequently $H_2SO_4$ as outlined above; as well as by the two reactions :

$H_2SO_4 + 2NaCl \rightarrow Na_2SO_4 + 2H^+ + 2Cl^-$ ...(27)

and $2NH_4^+ + SO_4^{2-} + 2H_2O \rightarrow 2NH_3 + 2H_3O^+ + SO_4^{2-}$ ...(28)

Relatively pure rain water has a pH around 5, 5-5.7, but owing to $SO_2$ emmissions the pH of rain can drop as low as 2. This increases the acidity of waterways, in particular lakes. Some Scandanavian lakes receiving rain, in which the $H_2SO_4$ aerosol originated in Europe and the U.K. have a pH of 5.5 to 4. Three quarters of the sulphur precipitated in Scandanavi originates from Europe and the U.K. In addition to increasing acidity of lakes and effecting aquatic like, acid rain accelerates leaching of nutrients from soils, affects the metabolism of soil organisms, increases metallic corrosion and destruction of basic building materials such as limestone and marble.

The acidity of natural water systems also increases by the dissolution of $SO_2$. Approximately 4 to 8 x $10^7$ tonnes of $SO_2$ dissolve in the oceans annually.

# 10

# Particulates

**Introduction and General Properties**

Small solid particles and liquid droplets, collectively called particulates, are present in the air in great numbers, and at times they constitute a serious pollution problem.

Many different chemical substances can enter the atmosphere in particulate form. The wide range in chemical composition of products from a single source, coal combustion, is shown in Table 1.

**Table 1. Fly Ash Composition (Coal Combustion)**

| *Component* | *Percentage of Fly Ash (% Calculated As)* |
|---|---|
| Carbon | 0.37-36.2 |
| Iron (as $Fe_2O_3$ or $Fe_3O_4$ | 2.0-26.8 |
| Magnesium (as MgO) | 0.06-4.77 |
| Calcium (as CaO) | 0.12-14.73 |
| Aluminium (as $Al_2O_3$) | 9.81-58.4 |
| Sulfur (as $SO_2$) | 0.12-24.33 |
| Titanium (as $TiO_2$) | 0-2.8 |
| Carbonate (as $CO_3$) | 0-2.6 |
| Silicon (as $SiO_2$) | 17.3-63.6 |
| Phosphorus (as $P_2O_2$) | 0.07-47.2 |
| Potassium (as $K_2O$) | 2.8-3 0 |
| Sodium (as $Na_2O$) | 0.2-09 |
| Undetermined | 0.08-18.9 |

Due to the large variety substances encountered, a general discussion of chemical properties cannot be made for particulate pollutants.

Many physical properties may apply to particulates in general. The most important of these is size, which ranges from a diameter of 0.0002 m (about the size of a small molecule) to a diameter of about 500 m. The micron (mh) is a commonly used unit in particulate pollution work and is equal to $10^{-6}$ metres. In the size ranges given above, particles are having lifetime in the suspended state of something between a few seconds and several months. This lifetime depends upon the settling rate, which in turn depends upon the size and density of the particles and the turbulence of the air.

A second property of particulates is their ability to act as sites for sorption. This property has been a function of surface area, which is large for most particulates. The process of sorption occurs when an individual molecule impacts on the surface of a particulate and does not rebound, but sticks or sorbs.

Obviously the optical properties of particles are exerting an important effect upon air quality. Particles less than approximately 0.1m in diameter scatter light much like molecules, i.e., Rayleigh scattering. Generally, such small particles are exerting an in significant effect upon visibility in the atmosphere. The light scattering and intercepting properties of particles larger than 1m are almost proportional to the particle cross-sectional area. Particles having diameters between 0.1m and 1m are of sizes having the same order of magnitude as the wavelengths of visible light. The optical properties are important in the determination of effects of atmospheric particulates on solar radiation and visibility.

Very small solid particles are carbon black, silver iodide, combustion nuclei, and sea salt nuclei. Larger particles are cement dust, windblown soil dust, foundry dust, and pulverized coal. Liquid particulate matter, generally categorized as "mist", includes rain drops, fog, and sulfuric acid mist. Some particulate matter is biological material, such as very small viruses, bacteria, bacterial spores, fungal spores, and pollen. Particulate matter may be organic or inorganic; both types have been very important atmospheric contaminants.

Particulate matter originates through a wide variety of processes which are ranging from simple grinding of bulk matter to complicated chemical or biochemical syntheses. The effects of particulate matter are also widely varied. Either by itself, or in combination with gaseous pollutants, particulate matter may be detrimental to human health. Atmospheric particles are able to damage materials reduce visibility, and bring about undesirable esthetic effects.

**Concentration of Particulates**

The units used to express concentrations of particulates are micrograms per cubic metre of air (mg/m). In order to convert from mg/m$^3$ to ppm on a volume basis, the molecular weight of the particulate is needed. No acceptable particulates molecular weights are available because of the diverse composition of particulates.

The number of particles in the atmosphere have been found to vary from several hundred per cubic centimetre in ultra clean air to more than 100,000 per cubic centimeter in highly polluted atmospheres. Most of the particulate mass in the atmosphere occurs in the size range of 0.1 m to 10 m. In remote areas the levels of atmospheric particulate mass may be as low as 10 mg/m$^3$. In urban areas average levels of 60 mg/m$^3$ to 220 mg/m$^3$ are generally encountered, and levels may range up to the vicinity of 2000 mg/m$^3$.

The size and composition of particulate matter have been found to vary diurnally and with weather conditions. Furthermore, the elemental composition differs markedly with particle size, and this may exert a strong influence upon the toxicological effects of particulate matter. For example, calcium tends to occur in large particles which are produced from wind erosion of soil, rock quarrie's cement, and similar sources. Likewise, the common soil components—iron, manganese, potassium and titanium—generally occur in large particles. Lead and bromine occur to a large extent in smaller particles, around the 0.5 m size. A particular toxicological hazard is presented, because particles in the range of 0.015-1 m have been most efficiently retained in the respiratory system. These two elements are found together in atmospheric particulate matter and get produced to a large extent by automobile exhausts.

**Sources of Particulates**

Numerous natural processes release particulate matter into the atmosphere. Typical of these processes are volcano eruptions and the blowing of dust and soil by the wind. The activities of man also release; for example, in the form of dust and absestos particulates from construction, fly ash from smelters and mining operations, and smoke from incomplete combustion processes.

**Physical Processes for Particle Formation**

The particles above approximately 1m in size are formed by the disintegration of larger particles. This is a dispersion process, and the product is termed as a dispersion aerosol. Dusts are solid dispersion aerosols.

Many dispersion aerosols originate from natural sources. Sea spray, windblown dust, and volcanic dust of course preceded man. Cultivation of land has made it much easier for the wind to pick up topsoil and blow it over great distances.

Much are more energy is needed to break material down into small particles than is needed (or released) to synthesize particles, through adhesion of smaller particles or through chemical processes. That is the reason that most dispersion particles are relatively large. These larger particles in general having fewer harmful effects than do very small particles.

**Chemical Processes for Particle Formaiton**

Metal oxides form a major class of inorganic particles in the atmosphere. These are produced whenever fuels containing metals are burned. For example, during the combustion of pyrite-containing coal,

$$3FeS_2 + 8O_2 \rightarrow Fe_3O_4 + 6SO_2 \qquad ...(1)$$

particulate iron oxide is formed. Organic vanadium in residual fuel oil gets converted to particulate vanadium oxide. Part of the calcium carbonate in the ash fraction of coal gets converted to calcium oxide,

$$CaCO_3 + heat \rightarrow CaO + CO_2 \qquad ..(2)$$

and goes to the atmosphere through the stack.

A special case of particulate matter formed during the combustion of fuel is that of the particulate lead halides which are formed in the burning of leaded gasoline. Tetraethyl lead in leaded gasoline reacts with oxygen and halogenated scavengers,

$$Pb(C_2H_4)_4 + O_2 + halogenated \rightarrow scavengers$$

$$CO_2 + H_2O + PbCl_2 + PbBrCl + PbBr_2 \qquad ...(3)$$

to form lead halides which are volatile enough to exit through the exhaust system, but which condense to form particles. By this pathway some lead has been entering the atmosphere.

A common process for the formation of aerosol mists involves the oxidation of atmospheric sulfur dioxide to sulfuric acid,

$$2SO_2 + O_2 + 2H_2O \rightarrow 2H_2SO_4 \quad ...(4)$$

a hygroscopic substance which accumulates atmospheric water to form small liquid droplets. In the presence of basic air pollutants, such as ammonia or calcium oxide, the sulfuric acid reacts to form salts:

$$H_2SO_4\ (droplet) + 2NH_3(gas) = (NH_4)_2SO_4(droplet) ...(5)$$

$$H_2SO_4(droplet) + CaO(particle) = CaSO_4(droplet) + H_2O. \quad ...(6)$$

Under conditions of low humidity water gets lost from these droplets and a solid aerosol is formed.

**Surface Properties and Reactions of Particles**

The three most important surface properties of particles include nucleation, adhesion, and sorption. Nucleation is involved in raindrop formation.

Adhesion, or coagulation, refers to the process by which extremely small particles form larger aggregates and eventually reach a size at which they settle rapidly. Liquid particles of the same composition are very likely to adhere to one another when they collide. The likelihood of solid particles adhering to one another tend to increase with decreasing particle size. Sorption designates the phenomenon by which molecules are taken up particles. If the vapour or gas is dissolved within the particle, the phenomenon is termed as absorption. Adsorption refers to sorption on the surface of the particle. Chemisorption refers to sorption involving a specific chemical interaction like the reaction of atmospheric $CO_2$ with particles of $Ca(OH)_2$:

$$Ca(OH)_2 + CO_2 \rightarrow CaCO_3 + H_2O$$

Other examples of chemisorption include the reactions of sulfur dioxide with aluminium oxide or iron oxide aerosols and the reaction of sulfuric acid aerosols and ammonia.

**The Composition of Inorganic particulate Matter**

In general, the proportions of elements in atmospheric particulate matter reflect the relative abundances of elements in the parent material. This provides a good idea of the source of much particulate matter. For example, particulate matter largely from ocean spray origin in a coastal area receiving sulfur dioxide pollution may show anomalously high sulfate and correspondingly low chloride. The sulfate comes from atmospheric oxidation of sulfur dioxide to form nonvolatile ionic sulfate, whereas some chloride may get lost from the solid aerosol as volatile HCl:

$$2SO_2 + O_2 + 2H_2O = 2H_2SO_4 \quad \text{...(1)}$$

$$H_2SO_4 = 2NaCl \text{ (particulate)} = Na_2SO_4 \text{ (particulate)} + 2HCl \text{ (gas)} \quad \text{...(2)}$$

Atmospheric particulate matter is having an extremely wide diversity of chemical composition. Organic matter, nitrogen compounds, sulfur compounds, a number of metals, and radionuclides occur in the particulate matter in polluted urban atmospheres. An extensive analysis of particulate matter collected from a large number of sampling stations in urban locations showed the following composition of particulate matter in units of $mg/m^3$ (arithmetic averages of biweekly samples): total suspended particulates (105), ammonium (1.3), nitrate (2.6), sulfate (10.6), benzene-soluble organics (6.8), antimony (0.01), arsenic (0.02), cadmium (0.002), chromium (0.015), copper (0.09), iron (1.58), lead (0.79), manganese (0.10), nikel (0.034), tin (0.02), titanium (0.04), vanadium (0.050), and zinc (0.67). Molybdenum occurred at an average level of > 0.005 mg/m$^3$ although a maximum value of 0.78 mg/m was observed. The average levels of beryllium, bismuth, and cobalt found were all less than 0.0005 mg/m$^3$ although maximum values of 0.010, 0.064, and 0.060 m/m$^3$, respectively, were observed. Although it is not included in the above list, sodium is a common element found in particulate matter in coastal areas. Sodium enters the atmosphere as sodium chloride in sea spray aerosol particles. Levels of approximately 1 mg/m$^3$ of sodium are not uncommon in coastal areas. Radionuclides occurred in the particulate matter, with a gross beta radio activity of 0.8 pico Curies/m$^3$, and a maximum value of 12.4 pCi/m$^3$.

At various times of the year or day and in different locations any of the trace elements which are of concern for their toxicological effects are found in atmospheric particles. Table 2 shows some elements found in particulate matter collected at Tucoson, Arizona.

**Table 2. Trace Elements from a Typical Particulate Sample**

| *Element* | *Concentration* mg/m$^3$ of air | *Element* | *Concentration* mg/m$^3$ of air |
|---|---|---|---|
| Al | 4.1 | Li | 0.003 |
| Be | 0.00003 | Mg | 1.3 |
| Bi | 0.003 | Mn | 0.012 |
| Ca | 10.4 | Na | 2.2 |
| Cd | 0.020 | Ni | 0.013 |
| Co | 0.0015 | Pb | 0.8 |
| Cr | 0.004 | Rb | 0.003 |
| Cs | 0.0004 | Si | 13.4 |

| | | | |
|---|---|---|---|
| Cu | 0.6 | Sr | 0.022 |
| Fe | 3.3 | Ti | 0.15 |
| K | 2.8 | V | 0.009 |
| | | Zn | 0.14 |

**Toxic Metals in the Atmosphere**

Some of the metals found in polluted atmospheres are hazardous to human health. All except beryllium belong to the classification of heavy metals. The TLVs of metals have been compared with their concentrations in a number of urban air samples. The results are summarized in Table 3.

**Table 3. Toxic Metals in the Atmosphere Compared to TLV and Average Levels Found in Urban Atmospheres**

| *Metal (descending order of toxicity* | *TLV mg/m³ (range denotes different toxic forms)* | *Average metal level: TLV* |
|---|---|---|
| Be | 0.002 | |
| less than 0.00025 | | |
| Hg | 0.01—0.05 | |
| 0.002 | | |
| Cd | 0.1 | |
| 0.0001 | | |
| Pb | 0.1—0.2 | |
| 0.004 | | |
| As | 0.2—0.5 | |
| 0.00004 | | |
| V | 0.5 | |
| 0.0002 | | |
| Ni | 0.007—1.0 | |
| 0.00003 | | |
| Cu | 1.0 | |
| 0.00009 | | |
| Fe | 1.0 | |
| 0.0003 | | |
| Zn | 1.0 | |
| 0.0001 | | |

**Mercury in the Atmosphere**

Mercury in the atmosphere has been a cause of great concern due to its toxicity. Elemental mercury is volatile, and much of the mercury entering the atmosphere does so in atomic form. Most of the mercury entering the atmosphere from coal combustion and volcanoes does so as the elemental form.

Some mercury in the atmosphere gets associated with particulate matter. Volatile organomercury compounds such as dimethylmercury, $(CH_3)_2Hg$, also occur in the atmosphere. Monomethylmercury salts, such as $(CH_3)HgBr$, are also volatile mercury compounds which sometimes occur in the atmosphere. Most coals are having appreciable levels of mercury, and this presents a problem which must be dealt with as the use of coal inevitably increases.

**Lead in the Atmosphere**

Lead is probably the most serious atmospheric heavy metal pollutant. Deecreasing use of leaded gasoline and increasing use of coal will probably result in its replacement by mercury as the most troublesome atmospheric metal. The lead halides from automotive exhausts are the most common form of atmospheric lead. Of these, lead bromochloride PbBrCl, predominates. Appreciable amounts of $NH_4Cl.2PbBrCl$, and ammonium chloride-lead bromochloride double salt, also occur in auto exhausts.

The atmospheric reactions and ultimate fate of this lead are subject to conflicting opinions.

**Beryllium in the Atmosphere**

Beryllium is used primarily as a component of specially alloys. Some of the primary uses include electrical equipment, electronic instrumentation, space gear, and nuclear equipment. Its distribution is by no means comparable to that of other toxic metals such as lead or mercury. Because of its specialty uses, however, consumption of beryllium is increasing. Beryllium has the lowest allowable TLV of any metal, and its presence in the atmosphere is very hazardous.

**Asbestos in the Atmosphere**

Asbestos is a naturally-occurring fibrous silicate. The fibers are separable, yielding fine filaments. Small fibers of asbestos irritate lung tissue causing a condition called "asbestosis". This condition is characterized by lung fibrosis. Therefore, atmospheric asbestos is a matter of considerable concern.

**Mineral Particulate Matter from Combustion—Fly Ash**

Much of the mineral Particulate matter in a polluted atmosphere is in the form of oxides and other compounds produced during the combustion of high-ash fossil fuel. Most of the mineral matter in fossil fuels such as coal or lignite is converted during combustion to a fused, glassy bottom ash which presents no air pollution problems. Smaller ash particles called fly ash enter urnace flues and, in a properly equipped stack system, are collected efficiently. However, a certain percentage of this fly ash escapes through the stack and enters the atmosphere. Unfortunately, these tend to be the smaller particles which do the most damage to human health, plants, and visibility.

**Radionuclides in the Atmosphere**

A significant natural source of radionuclides in the atmosphere is the noble gas radon, product of the decay of radium. Radon may enter the atmosphere as one of two isotopes. $^{222}Rn$ and $^{220}Rn$. Both are alpha emitters with half-lives of 3.8 days and 54.5 seconds, respectively. Also, cosmic rays act on nuclei in the atmosphere to produce radionuclides including $^{7}Be$, $^{10}Be$, $^{14}C$, $^{39}Cl$, $^{3}H$, $^{22}Na$, $^{32}P$ and $^{33}P$.

One of the more serious problems in connection with radon appears to be that of radioactivity originating from uranium mill tailings. Some medical authorities have postulated that the rate of birth defects and infant cancer in areas where uranium mill tailings have been employed in residential construction are significantly higher than normal.

The combustion of fossil fuels introduces radoactivity into the atmosphere in the form of radionuclides contained in fly ash. Large coalfired power plants lacking ash control equipment may introduce up to several hundred milicicuries of radionuclides into the atmosphere each year.

The radioactive noble gas $^{35}Kr$ goes into the atmosphere by the operation of nuclear reactors and the processing of spent reactor fuels.

The detonation of nuclear weapons above ground contributes a great deal to particulate radioactive material. Such material may be carried through the atmosphere for tremendous distances, eventually to be scavenged from the atmosphere by rain. For example, about a week after a nuclear detonation in China on December 24, 1967, the following radionuclides were found in rain in Fayetteville, Arkansas: $^{91}Y$, $^{141}Ce$, $^{144}Ce$, $^{147}Nd$, $^{147}Pm$, $^{149}Pm$ $^{151}Sm$, $^{153}Sm$, $^{155}Eu$, $^{89}Sr$, $^{90}Sr$, $^{115m}Cd$, $^{129m}Te$, $^{131}I$, $^{132}Te$, $^{140}Ba$. The rate of travel of radioactive particle through the atmosphere is a function of particle size. Appreciables fractionation of nuclear debris is observed because of the difference in rate of transport.

**Organic Particulate Matter**

Organic particulate matter is found in a wide variety

of compounds in the atmosphere. It is interesting to note that a so called "average formula" of this organic matter, collected from 200 U.S. areas and isolated as a benzene-extractable fraction, is $C_{32.4}H_{48}\ O_{3.8}S_{0.083}$ $(\text{hologen})_{0.065}$ $(\text{alkoxy})_{0.12}$. Much of this organic matter occurs in the respirable 1m size range.

The aliphatic fraction of the neutral group is having a high percentage of long-chain hydrocarbons, predominantly those containing from 16 to 28 carbon atoms. Such hydrocarbons are not though to be particularly significant health hazard, or to participate strongly in atmospheric chemistry reactions. Many of the hydrocarbons found in the aromatic fraction, however, are polynuclear aromatics, and some are proven carcinogens (cancer-causing agents). Therefore, these hydrocarbons are particularly important as air pollutants. Aldehydes, ketones, epoxides, esters, quinones, and lactones are found among the oxygenated neutral organic components. A few of these compounds may be mutagenic or carcinogenic. The acidic fractions are having long-chain fatty acids and nonvolatile phpenols. Among the acids recovered from air pollutant particulate matter are lauric, myristic, palmitic, stearic, behenic, oleic, and linoleic. The basic fraction consists largely of alkaline N-heterocyclic hydrocarbons, such as acridine:

At the present time comparatively little is known about the chemical constitution and environmental significance of the basic fraction.

A significant portion of organic particulate matter coming from automotive exhausts has been characterized as "oxidized, polymerized hydrocarbons and nitrogenous azaheterocyclic substances".

**Table 4. Types of Organic Compounds Found in Respirable Organic Matter along Highways.**

| *Type of compound* | *Possible effect* |
|---|---|
| Polar organics | Extracted by aqueous body fluids enabling penetration of tissue in a soluble form. |
| Aldehydes | Suspected of being the primary causative agents in smoke poisoning, postulated denaturation of amino acids and RNA. This causes inflamation and necrosis of tissue. |
| Other oxygenated organics | Some have suspected carcino |
| (acids, neutral oxygen compounds) | genic effects. |
| Polynuclear azahetero | Some of these, such as dibenz |
| cyclinc compounds (benza | (a, h)-acridine and dibenz (a, j) |
| cridines, dibenzacridines, | acridine are known carcinogens. |
| benzoquinolones) | |

**Effects of Particulates on Plants**

Relatively little research has been done on the effects of particulate pollution on vegetation. Most of what has been done has been concerned with specific dusts rather than the conglomerate mixture usually present in the atmosphere. This dust, when combined with a mist or light rain, formed a thick crust on upper leaf-surfaces which was found to interfere with photosynthesis in the plant by shielding out needed sunlight and upsetting the process of $CO_2$ exchange with the atmosphere. The growth of such plants was inhibited.

A possible indirect effect of particulates deposited on plants is the chance that the particulates might contain chemical components harmful to animals that use the plants as food.

**Effects of Particulate on Humans**

The only major route by which particles enter the body through the respiratory tract. Relatively large particles are likely to get retained in the nasal cavity and in the pharynx. Very small particles are likely to reach the lungs and to be retained by them. The respiratory system is having mechanisms for the removal of particulate matter. In the ciliated region, particles are carried as far as the entrance to the gastrointestinal tract by a flow of mucus. Macrophages in the nonciliated pulmonary regions carry particles to the ciliated region.

The respiratory system may get damaged directly by particulate material. It may enter the blood system or lymph system through the lungs. In addition, the particulate material or soluble components of it may get transported to organs some distance from the lungs and have a detrimental effect on these organs. Particles cleared from the respiratory tract largely enter the gastrointestinal tract. Material may also enter the blood after it is transported to the gastrointestinal tract.

A strong correlation exists between the daily mortality rate and acute episodes of air pollution. In such cases, high levels of particulate matter are accompanied by high levels of $SO_2$ and other pollutants.

In Fig. 2 and 3 the main parts of the respiratory system are given, including a detail of the terminal pulmonary structure with its tiny alveolar (air) sacs. These alveolar sacs, resulting from much branching of the air ducts, are the sites at which oxygen and carbon dioxide are exchanged between the atmosphere and the blood.

Particles over 5.0 μ in diameter are stopped and deposited mainly in the nose and the throat. Those that do penetrate to the lungs do not often go beyond the air ducts or bronchi, and even these are soon removed by ciliary action. Particles ranging in size from 0.5—5.0 m in diameter can be deposited in lung as far as the bronchioles. Few reach the alveoli. Most particles deposited in the bronchioles are removed by the cilia within the two hours. Particles less than 0.5 in diameter reach and may settle in the alveoli. The removal of such particles from these areas is much less rapid and less complete than from the larger passages. Some of the particles retained in the alveoli are absorbed into the blood.

Particulate matter which enters and remains in the lungs can exert a toxic effect in three different ways.

1. The particles may be intrinsically toxic because of chemical or physical characteristics. More will be said later about this type of particle.
2. The particles may be inert themselves but once in the respiratory tract they may interfere with the removal of other more harmful material.
3. The particles may carry adsorbed or absorbed irritating gas molecules and thus enable such molecules to reach and remain in sensitive areas of the lungs. Carbon, in the form of soot, is a common particulate with a very good ability to sorb gas molecules on the surface.

**Toxic Particulates**

Intrinsically toxic particulates are not commonly found

in the atmosphere in high concentrations except for sulfuric acid aerosol. However, many toxic particles are found in the air in trace amounts.

**Table 5. Trace Metals that may pose Health Hazards in the Environment**

| *Element* | *Sources* | *Health Effects* |
|---|---|---|
| Nickel | Diesel oil, residual oil, coal, tobacco smoke, chemicals and catalysts, steel and nonferrous alloys | Lung cancer (as carbonyl) |
| Beryllium | Coal, industry (new uses proporsed in nuclear power industry, as rocket fuel) | Acute and chronic syste poison, cancer |
| Boron | Coal, cleaning agents, medicinals, glass making, other industrial | Nontoxic except as borane |
| *Element* | *Sources* | *health Effects* |
| Germanium | Coal | Little innate toxicity |
| Arsenic | Coal, petroleum, detergents, pesticides, mine tailings | Hazard disputed, may cause cancer |
| Selenium | Coal, sulfur | May cause dental caries, carcinogenic in rats, essential to mammals in low doses |
| Yttrium | Coal, petroleum | Carcinogenic in mice over lon-term |
| exposure | | |
| Mercury | Coal, electrical batteries, other industrial | Nerve damage and death |
| Vanadium | Petroleum (Venezuela, Iran), chemicals and catalysts, steel and nonferrous alloys | Probably no hazard at current levels |
| | | *Contd.* |
| Cadmium | Coal, zinc mining, water mains and | Cardiovascular disease and hypertension |

| | pipes, tobacco smoke | in humans suspected interferes with zinc and copper metabolism |
|---|---|---|
| Antimony | Industry | Shortened life span in rats. |
| Lead | Auto exhaust (from gasoline), paints (prior to about 1948) | Brain damage, convulsions, behavioural disorders, death. |

The poisonous nature of beryllium and its compounds has been recognized for years. Soluble compounds such as $BeSO_4$ and $BeCl_2$ commonly produce acute inflammation of the lungs. Beryllium metal or insoluble compounds cause the chronic pulmonary disease, berylliosis.

Coal burning is another source of atmospheric beryllium because most coal contains beryllium as an impurity.

Recent findings indicate that inhalation of asbestos, a substance used for insulation by the building trade, can produce serious problems in previously healthy lungs.

**Table 6. Beryllium Exposure Limits Set by the AEC**

| *AEC Exposure Limit (mg/m³)* | *Conditions Levels mg/m³)* | *Measured Atmospheric* | *Conditions* |
|---|---|---|---|
| 25 | Industrial limit for short-term exposure | 0.028—0.083 | Concentration in vicinity of a beryllium plant |
| 2. | Industrial limit averaged an eight hour work period | 0.008 | Maximum ambient air value |
| 0.01 | Industrial limit averaged over 30 | 0.0005 | Daily average air value |

| | days for people living in the neighbourhood of a beryllium plant | |
|---|---|---|

**Effects of Particulates on Materials**

Airborne particles, including soot, dust, fumes, and mist can bring about a wide range of damage to materials. The extent and type of damage have been found to depend upon the chemical composition and physical state of the pollutant. Passive damage is caused when particulates settle on and soil materials, creating a need for more frequent cleaning. The cleaning processes weaken the materials. Chemical damage results when the particulates themselves are corrosive, or when they carry absorbed or absorbed corrosive substances along with them.

Metals are generally resistant to corrosion in dry air and even in clean air that are having only small amounts of water. Particulates accelerate corrosion, especially in the presence of sulfur-containing compounds.

The ability of particulates to damage and soil buildings, sculpture, and other structures is especially more common in cites where large amounts of coal and sulfur-bearing oils are used. The resulting particulates are generally tarry, sticky, and acidic so they adhere to surfaces and behave as acid reservoirs for corrosion.

Painted surfaces are very susceptible to particulate damage before the paint is dry. In addition, some particulate fumes and mists react directly with dry painted surfaces, and bring about considerable damage. This type of paint damage is common on automobiles frequently parked near industrial plants.

The soiling of textile materials has been one of the most important harmful effects from particulates.

**Effects of Particulates on Solar Radiation and Climate**

Particulates in the atmosphere are known to exert a definite influence on the amount and type of solar radiation that reaches the surface of the earth. This influence is the result of light scattering and absorption by the particulates. One principal effect is the reduction of visibility. The eye has difficulty in distinguishing between the object and the background. Decreased visibility creates obvious problems, some dangerous (aircraft or automobile operation) and some just annoying (sightseeing, and so forth). Illumination problems are created by particulate pollution. In some cities the processes of scattering and absorption are able to reduce by one-third or more the amount of visible light reaching the earth's surface. The gloom resulting from reduced amounts of sunlight creates a heavy need for artificial illumination, which in turn require more power which could mean that more pollutants from the power plant will be released, and a cycle is created.

The amount of particulate pollution tend to vary with the season. In autumn and winter, increased heating of houses, and other buildings creates a demand for mor power which causes more particulates. This perhaps accounts in part for gloomy winter day.

Particulates in the atmosphere may influence the formation of clouds, rain, and snow by acting as nuclei upon which water condensation can take place.

World-wide climate changes are also related to particulate pollution. The reduction in the amount of solar radiation reaching the earth's surface may be upsetting the delicate heat balance of the earth's atmosphere. In spite of an increase in $CO_2$ content of the atmosphere over the past several decades, which should cause an atmospheric temperature increase, the average world-wide temperature has been decreasing slightly since 1940s. The increased reflection of solar radiation by particulates has been more than compensating for the climatic effect of increased atmospheric $CO_2$.

**Effect of Hydrogen Fluoride**

Hydrogen fluoride behaves somewhat similar to sulphur dioxide except that with a few species of plants it is effective in causing lesions and interfering with photosynthesis in concentrations two or three orders less than in the case of sulphur dioxide. With most species it is up to 10 times as effective as sulphur dioxide. However, recovery of plants from the fluoride effect is much slower than from sulphur dioxide. This difference in rate of recovery is probably explained by the fact that sulphite in the leaves is rapidly oxidised to the relatively non-toxic sulphate, whereas fluorides can be removed only by the slower process of volatization or by some obscure chemical reaction. Forage may be rendered unsafe for animal feeding if more than 50 ppm of fluorine is absorbed.

**Effect of Ozone**

Ozone is phytotoxic in exposures of a few hours at about 0.2 ppm. Injury due to ozone is quite different from typical smog injury. The lesions are generally confined to the upper surface.

**Fluorine**

Of all farm animals, cattle and sheep are the most suceptible to fluorine toxicosis. Horses appear to be quite resistant to fluorine poisoning. Poultry are probably the most resistant to fluorine, of all farm animals.

**Symptoms of Acute Fluorine Poisoning**

Lack of appetite, rapid loss in weight, decline in heath and vigour, lameness, periodic diarrhoea, muscular weakness and death. Characterize the acute form of fluorine poisoning. It may also result in considerable increase of bone fluorine. But acute poisoning due to fluorides is unlikely in a majority of the cases.

**Symptoms of Chronic Fluorine Poisoning**

Fluorine is cumulative poison under conditions of continuous exposure to sub-acute doses. Fluorine is also a protoplasmic poison. It has a marked affinity for calcium and interferes with normal calcification. Animals have been reported to be more resistant than humans to dental mottling. Cattle and sheep are the most frequently affected animals. Teeth in the process of formation are easily affected. Hence tooth symptoms are a sensitive and unique criterion of chronic fluorosis. Excessive wearing of incisor and molar teeth may occur at high levels of fluorine intake.

Bone lesions may develop at any age. A bony overgrowth may be observed on the leg bones, jaw bones and ribs. Their appearance indicates there has been a high fluoride intake over a long period. Lameness may occur as a result of this overgrowth on the leg bones. This may be followed by stiffness of joints. Effects of fluorides on skeletal structures are usually of a permanent nature.

Symptoms of advanced fluorisis include lack of appetite, general ill-health due to malnutrition, lowered fertility, reduced milk production and growth retardation. The symptoms developed in other species such as rabbits, horses and poultry are similar to those in cattle and sheep, i.e., mottling, staining and wearing of the teeth, bony overgrowths on the skeleton, stiffiness and lethargy and general ill-health from malnutrition and starvation.

While diagnosis of chronic fluorine poisoning is done, one has to be careful. This is because drinking water supplies may contain unduly high level of fluorides. Also, mineral supplements may contribute appreciable amounts of fluorine to the diet. Hence, a check of all possible sources must be made when investigating fluorosis.

**Arsenic**

Arsenic occurs as an impurity in many ores and in coal. It has been reported to cause poisoning of livestock near various industrial processes and smelters. Just like most industrial air contaminants, arsenic may spread over a considerable area from a stack source. Arsenic in dusts or sprays on plants can lead to poisoning of cattle.

**Symptoms of Acute Arsenic Poisoning**

In acute cases the symptoms are severe salivation, thirst, vomiting, uneasiness, feeble and irregular pulse and respiration. The animal stamps, lies down and gets up. There is diarrhoea, and the faeces have a garlic odour and are sometimes bloody. The ears become cold, the body trembles and develops abnormal temperature and convulsion. Death may occur in few hours or days.

**Symptoms of Chronic Arsenic Poisonfng**

Arsenic appears to have a depressing effect upon the central nervous systems. The animal becomes dull, and exhibits a lack of appetite, with a resulting weight loss. Also, there may be chronic cough and diarrhoea may occur continuously. There may be thickening of the skin, anaemia and abortion or sterlity. Chronic poisoning can result in eventual paralysis and death.

**Tolerance Limits**

Arsenic as well as its soluble compounds are extremely poisonous. It has been reported that sheep have been poisoned by as little as 0.25-0.50 g of arsenic daily, whereas cattle and horses may tolerate 1.3-1.9 g daily. It has also been reported from laboratory experiments, that 10 mg of arsenic per kilogram of body weight per day has produced chronic arsenic poisoning in guinea pigs. A detailed investigation of the effects of arsenic poisoning on farm animals is necessary to establish standard tolerance limits.

**Lead**

Lead contamination of the atmosphere takes place on account of various industrial sources such as smelters, coke ovens and other coal combustion processes. Lead is also used in dust and sprays containing lead arsenate.

**Symptoms of Acute Lead Poisoning**

In case of acute lead poisoning the onset is sudden and the course relatively short. Prostration, staggering and inability to rise are prominent symptoms. The pulse is always fast but weak. Some animals may fall suddenly, stiffen the legs and have convulsions. There is complete loss of appetite, paralysis of the digestive tract and diarrhoea. Other nervous symptoms in cattle are grinding of the teeth and rapid chewing of the cud.

In case of horse, it may result in complete loss of appetite, nervous depression, lethargy and death.

**Symptoms of Chronic Lead Poisoning**

Chronic lead poisoning has been observed frequently in horses that have been grazing an forage near smelters, lead mines, and in orchards that have been sprayed. Paralysis of the muscles of the larynx and difficulty in breathing are the main symptoms. Convulsions may occur from the paralysis of the throat, and the difficulty in breathing may be unusually severe and persistent, during and after exercise.

**Tolerance Limits**

Lead is a cumulative poison. Therefore, the continuous ingestion of even very small daily doses will be finally as effective as one toixc dose. Hence in case of slight contamination death can occur after many months and in case of severe contamination death can occur within 24 hours.

It has also been reported that inadequate calcium in diet increases the retention of lead in the animal body. Low calcium diets may result in lead storage of as much as five times as compared to that found in animals that receive sufficient amounts of calcium. However, extra quantities of calcium above the amount considered sufficient, will not offer additional protection against poisoning.

**Hydrogen Sulphide and Mercaptans**

Hydrogen sulphide is a foul smelling gas. It is well known for its rotten egg like odour. Exposures to hydrogen sulphide for short periods can result in fatigue of the sense of smell.

Other sulphur compounds that are of interest in air pollution mainly because of their strong odours are methyl mercaptan ($CH_3SH$) and ethyl mercaptan ($C_2H_3SH$). But it has been reported that at the concentrations at which they are odour nuisances, they have no other effect on human health. In fact, mercaptans are often added to natural or manufactured gas supplies so that leakage of gas will be noticed.

**Fluorides**

Fluorides present in air, range from those which are extremely irritant and corrosive like hydrogen fluoride to relatively non-reactive compounds. But fluorine is a cumulative poison even under condition of prolonged exposure and in subacute concentrations.

**Lead**

The main source of lead in urban atmospheres is the automobile. It creates urban concentration of inorganic lead of about 103 mg/m$^3$, with high values in areas of heavy traffic. Inorganic lead acts as an agent which causes a variety of human health disorders. The effects include gastro-intestinal damage, liver and kidney damage, abnormalities in fertility and pregnancy, and mental development of children gets affected.

**Hydrogen Vapours**

Some of the hydrocarbon vapours in the atmosphere have health implications. The effect of formaldehyde is primarily irritating. It is a major contributor to eye and respiratory irritation caused by phtochemical smog.

**Carcinogenic Agents**

Carcinogenic agents are responsible for cancer. For example, the poly-cyclic organic compounds, 3, 4-benzpyrene. The origin of these compounds is in the incomplete combustion of hydrocarbons and other carbonaceous materials. They are also reported to be present in exhaust discharges from I.C. engines. In addition to poly-cyclic organic compounds, it has been found that some aliphatic hydrocarbons are also carcinogenic.

**Insecticides**

Insecticides are not only harmful for insects but also

poisonous for man, e.g., DDT (Dichloro diphenyl trichloroethane). They can affect the central nervous system and may attack other vital organs. In fact, DDT has been found in mother's milk in western countries and even in our own country. Hence, the use of DDT has been banned in the USA. The United Nations Environment programme (UNEP) has warned against the undesirable effects of indoor spraying of DDT. It has also been reported that indoor spraying effects domestic livestock.

# Index